# Maturity Business Models for Manufacturing in the Digital Age

**Bożena GAJDZIK**
Industrial Informatics Department
Silesian University of Technology, Poland

**Magdalena JACIOW**
Department of Digital Economy Research
University of Economics in Katowice, Poland

**Radosław WOLNIAK**
Organization and Management Department
Silesian University of Technology, Poland

**Robert WOLNY**
Department of Digital Economy Research
University of Economics in Katowice, Poland

CRC Press
Taylor & Francis Group
Boca Raton  London  New York

CRC Press is an imprint of the
Taylor & Francis Group, an **informa** business

A SCIENCE PUBLISHERS BOOK

First edition published 2025
by CRC Press
2385 NW Executive Center Drive, Suite 320, Boca Raton FL 33431

and by CRC Press
4 Park Square, Milton Park, Abingdon, Oxon, OX14 4RN



*Library of Congress Cataloging-in-Publication Data (applied for)*

ISBN: 978-1-032-72970-1 (hbk)
ISBN: 978-1-032-74168-0 (pbk)
ISBN: 978-1-003-46795-3 (ebk)

DOI: 10.1201/9781003467953

Typeset in Times New Roman
by Prime Publishing Services

# Preface

In an era where digitalization is revolutionizing industries, the quest for business maturity has never been more critical. This book provides insights, methodologies, and practical guidance to navigate the complexities of digital transformation. Authored by experts from the Silesian University of Technology and the University of Economics in Katowice, Poland—Bożena Gajdzik, Magdalena Jaciow, Radosław Wolniak, and Robert Wolny—the book distills years of research and practical experience into a coherent framework. These authors offer a holistic perspective on the maturity of manufacturing businesses in the digital age.

The book meticulously examines the concept of business maturity, emphasizing the readiness to adopt new technologies and adapt to changing market conditions. At its heart is the Industry 4.0 maturity model, a framework designed to assess an organization's readiness to implement advanced technologies, evaluating dimensions linked to key Industry 4.0 pillars such as the Internet of Things, Big Data, Cloud Computing, and Artificial Intelligence. This model serves as a vital tool for companies navigating the digital landscape.

Beginning with a historical perspective on industrial transformation, the book traces the evolution from traditional manufacturing to the digital age, setting the stage for a deeper exploration of business maturity models. Subsequent chapters delve into the conceptualization and operationalization of these models, offering detailed methodologies for measuring business maturity. Empirical research on Polish manufacturing companies forms a significant part of this work, providing a nuanced understanding of the current state of digital maturity, highlighting advancements and challenges. The practical implications are significant, offering recommendations for businesses to enhance their maturity levels and remain competitive in a rapidly evolving digital economy.

The book aims to equip managers, researchers, and industry practitioners with the knowledge and tools to guide their organizations through digital transformation, bridging theoretical concepts and practical applications.

# Contents

# Introduction

The concept of business maturity refers to an organization's ability to effectively manage its processes, structures, and resources to achieve strategic goals. The level of business maturity indicates how well a company is prepared to respond to changing market conditions, introduce innovations, and utilize modern technologies. Business maturity encompasses operational processes, Information Technologies (ITs) and tools, and management processes.

A mature organization has well-defined, automated, and optimized business processes that are constantly monitored and improved. This enables the company to respond quickly to changes and achieve high operational efficiency. Companies with a high level of business maturity effectively integrate modern technologies and IT tools into their operations. Business maturity also involves the ability to manage change effectively, whether due to internal or external factors such as new technology adoption, compliance with changing regulations, or market-trend responses. In mature organizations, an organizational culture focused on innovation, continuous improvement, and adaptability is observed. Leaders in such companies support employee development, promote collaboration, and are open to new ideas. Additionally, mature companies can effectively identify, assess, and manage risks. They have systems and processes in place to minimize the impact of negative events on operations. High business maturity is also associated with effective human resource management, which includes developing employee competencies, creating career paths, and fostering team commitment and motivation.

Assessing the business maturity of a company allows for the identification of areas needing improvement and the determination of development strategies that will enable the company to achieve higher efficiency and competitiveness. This is particularly important in the context of digital transformation, where the ability to quickly and effectively adapt to new technologies and changing market conditions is crucial for success.

Industry is entering a new era—the era of digitalization. *Manufacturing in the Digital Age: A Maturity Business Model for the Modern Industrial Transformation* is a book that aims to understand this transformative period in industrial production history. With technological advancements, traditional production methods are being revolutionized. Modern technologies not only increase efficiency and precision, but also spur innovations that were previously beyond the reach of enterprises.

In today's rapidly changing industrial environment, digital transformation is no longer optional but necessary. Companies that wish to remain competitive must not only implement new technologies, but also adapt their business models and organizational structures to new challenges. This book presents business maturity models describing companies at different stages of their digital journey.

The methodology adopted in this study for assessing business maturity levels is based on the functional goal of the designed research model, which is to evaluate the development stage of companies towards Industry 4.0. The concept of Industry 4.0 serves as the primary research paradigm in this model.

Industry 4.0 maturity is conceptualized as a holistic model assessing an organization's readiness to implement Industry 4.0 technologies. This model considers various dimensions linked to the key pillars of Industry 4.0 to gauge a company's technological progress as it transitions to an intelligent and interconnected ecosystem, thereby measuring its business maturity in the digital economy. The Industry 4.0 maturity model acts as a metric to assess the impact of transformation on business operations and maturity. It evaluates a company's maturity, helping identify strengths, weaknesses, and areas for improvement on the path to a technologically advanced future—a prerequisite for achieving full business maturity.

Business maturity is perceived as a progressive process characterized by distinct stages of development. Each stage defines criteria for a certain level of complexity and advancement. The created research model of the business maturity of manufacturing companies offers a framework and tools for not only assessing the current state (maturity stage) but also planning and executing transformation strategies, helping companies move from traditional production methods to advanced, digitally integrated processes.

The aim of this book is to provide readers with comprehensive knowledge about digital transformation in industries, guiding them towards business maturity in the context of Industry 4.0. The specific objectives are as follows:

1. Organizing concepts, definitions, and classifications related to business maturity models of industrial enterprises.
2. Conceptualizing and operationalizing the research model of business maturity of industrial enterprises.
3. Measuring business maturity of enterprises in accordance with the research model and evaluating the validity and reliability of the measurement.
4. Providing insight into the current state of business maturity in the context of the implementation of Industry 4.0 technologies in Polish manufacturing companies.
5. Identifying the challenges faced by companies implementing Industry 4.0 technologies.

6. Providing recommendations for companies to help them advance to a higher level of business maturity in the digital economy.

Measuring the maturity of manufacturing companies is crucial in the context of Industry 4.0 for several reasons. Maturity models provide a structured framework for assessing the current state of a company's technological and organizational capabilities. By assessing maturity, companies can identify their strengths and weaknesses, allowing them to prioritize areas for improvement and investment. This is especially important in Industry 4.0, where rapid technological advances and the integration of digital and physical systems require a clear understanding of a company's readiness to adopt new technologies.

Moreover, maturity models facilitate benchmarking against industry standards and competitors. By understanding their position relative to others, companies can set realistic goals and strategies for their digital transformation journeys. This competitive intelligence is essential to maintaining and strengthening market position in an increasingly digital and interconnected industrial environment. These models help align technological advances with business goals, ensuring that the adoption of new technologies such as the Internet of Things (IoT), Big Data, and Artificial Intelligence (AI) is strategically aligned to improve operational efficiency, product quality, and overall business performance.

Maturity models provide a roadmap for the systematic implementation of Industry 4.0 initiatives. They outline step-by-step progress and milestones that guide companies through the complexities of digital transformation, ensuring that changes are sustainable and effectively integrated into existing processes. Measuring maturity models support organizational change management, highlighting the need to upskill employees, adapt management practices, and foster a culture of continuous improvement and innovation. This holistic approach ensures that technological upgrades are complemented by human and organizational development, which is critical to the long-term success of Industry 4.0 initiatives.

The book consists of five chapters. Chapter 1 presents the conceptual framework for industrial transformation and the pathways to maturity. The chapter traces the evolutionary trajectory of the manufacturing sector from industrialization to the present day, with a particular focus on the shift towards digitalization and automated systems. It is divided into three parts: the historical perspective of industrial transformation, the framework for modern industrial transformation, and the steps taken by companies in the transformation process. The process of transforming enterprises towards Industry 4.0 is described step-by-step, starting from situation diagnosis, through digital and technological audits, to the implementation of changes within companies. The foundation of this process is digitization and automation, without which companies cannot initiate projects aimed at achieving Industry 4.0 standards. Finally, the chapter describes the pillars of Industry 4.0, highlighting key and emerging technologies.

Chapter 2 discusses a conceptual approach to business maturity models, highlighting their dynamic nature and role in facilitating organizational change and growth. The chapter explores Maturity Models (MMs) as structured sequences applicable to different entities, outlining the expected development path. Business process maturity, based on concepts such as Total Quality Management and Business Process Management, is studied as a measure of an organization's optimization and efficiency. It examines various MM applications, including digital transformation and organizational process maturity, and discusses the benefits and challenges of Business Maturity Models (BMM). Despite criticisms, the text emphasizes the practical value of these models as tools for a process-oriented approach, advocating empirical methodologies and stakeholder involvement in their development. The chapter also discusses the maturity of organizational processes, quality management systems, and technology readiness, highlighting models such as PEMM™ and ISO standards. It concludes with a detailed analysis of Industry 4.0 maturity models, focusing on frameworks such as SIMMI 4.0 and MEMM, and discussing the importance of strategy, collaboration, and leadership in addressing the challenges of the Fourth Industrial Revolution.

Chapter 3 delves into the concept of measuring business maturity. It begins with the conceptualization and operationalization of a research model aimed at assessing the business maturity of industrial enterprises. In the conceptual phase, the dimensions of business maturity are identified, and a research model is defined in line with the Industry 4.0 digital maturity model. In the operational phase, a digital maturity measurement model is created, covering the nine pillars of Industry 4.0: Internet of Things, Big Data, Cloud Computing, Advanced Simulation, Autonomous Systems, Universal Integration, Virtual and Augmented Reality, Additive Manufacturing, and Cybersecurity. The chapter discusses the methodological approach in detail, including the development of a tool for measuring business maturity and determining the reliability of the proposed measurement scale. The results, analyzed using statistical programs, show high reliability of the measurement scale. The research process is described in seven stages, from model creation to data analysis.

Chapter 4 presents the results of direct research conducted on a sample of 317 Polish industrial enterprises from the automotive, steel, and food industries. The choice of these industries for analysis is based on their unique characteristics and relevance to Industry 4.0. The automotive industry, with its high level of automation, extensive use of robotics, and significant investment in advanced manufacturing technologies, provides insight into cutting-edge technological implementations. The steel industry represents a more traditional sector facing efficiency, sustainability, and cost reduction challenges. The food industry, with its direct impact on consumer health and safety, and stringent regulatory requirements, benefits from Industry 4.0 technologies through increased traceability, improved quality assurance, and optimized supply chain management. The survey, conducted using an online survey technique supported by phone, took place between April 9 and May 6, 2024. Based on national databases, 1,508 companies were invited to participate, with a response rate of 20%, resulting in a final sample of 317 companies. The research, conducted in Polish, involved senior management respondents.

The collected data was analyzed using both simple and advanced statistical methods, focusing on the level of advancement in implementing Industry 4.0 technologies across these industries. The results provide a better understanding of which technologies are more or less advanced in the context of Industry 4.0 and how companies manage their implementation. The maturity models discussed offer insight into the different levels of technological maturity in various industries.

Chapter 5 provides a summary of the research on the business maturity of Polish manufacturing enterprises in the context of Industry 4.0. The chapter highlights key factors influencing technological advancement and differences in business maturity levels across the studied industrial sectors. It includes the results of a comparative analysis of Industry 4.0 maturity levels based on various characteristics of enterprises, such as industry type, employment size, year of establishment, and capital origin. The chapter identifies critical barriers to achieving higher levels of business maturity and discusses significant strategic implications for industrial enterprises, policymakers, and stakeholders in the digital economy. It also provides practical recommendations for companies to enhance their business maturity.

The book fills a gap in the knowledge of a holistic business maturity model specifically tailored to Industry 4.0. This model, covering nine key pillars, provides a structured framework for assessing and improving technology readiness. The robustness and reliability of the model are verified through empirical research, making it useful for both academic research and practical applications. While much of the existing literature on Industry 4.0 focuses on the global or Western European context, this study provides unique insights into the Polish manufacturing sector, contributing to a more nuanced understanding of regional differences in Industry 4.0 implementation.

An innovative approach in the book is the combination of multiple technological dimensions in a single maturity assessment. By considering various aspects of digital transformation—from advanced simulation to virtual reality and additive manufacturing—the study provides a comprehensive view that captures the complexity and interdependencies of modern manufacturing technologies. The practical implications of these findings are significant. The study not only identifies challenges and provides theoretical insights but also offers concrete recommendations and strategies for businesses and policymakers. This practical orientation makes the findings highly relevant and useful for real-world application.

This book is addressed to managers, scientists, researchers, and anyone interested in modern technologies and their application in industry. In an era of dynamic technological changes, knowledge about digital transformation is becoming a key element of success. We hope this publication will be a valuable source of information and inspiration for those guiding their organizations through the complex process of digitization, enabling them to fully leverage modern technologies' potential and remain competitive in the global market.

# 1 | Pathways to Modern Industrial Transformation

This chapter introduces the concept of industrial transformation along with the pathways to maturity. It traces the evolutionary trajectory of the manufacturing sector from the beginning of industrialization to the present day, with a particular focus on the transition toward digitization and automated systems. The chapter consists of three parts. The first section deals with the historical perspective of industrial transformation. The second section presents a framework for modern industrial transformation, and the third section deals with the steps that companies are taking in their transformation to Industry 4.0.

The transformation of companies to Industry 4.0 is presented step-by-step, from the diagnosis of the company's situation, to digital and technology audits, to projects in companies. The background of the transformation is digitization and automation. The conclusion provides a reference to the directions of Industry 5.0.

## 1.  Historical Perspective on Industrial Transformation

Digital transformation of industrial enterprises is a key element of the business strategies of many companies in the era of Industry 4.0 (I4.0). This transformation is a prerequisite for companies to participate in the development of the concept. The concept of "Industry 4.0" (in the German original: Industrie 4.0) was first introduced in a wide public space by Professor Klaus Schwab at the 2011 World Economic Forum Conference in Davos, and a group of politicians and scientists presented the concept at the Hannover Fair. Five years later, the government of the Federal Republic of Germany made the concept of Industry 4.0 an important direction for the development of the German industry. The assumptions of this concept have been set out in the government policy documents, including the Action Plan of the High-Tech Strategy 2020 adopted in 2014. On the basis of this strategy, specific projects—often in the form of public-private partnerships—have been launched in Germany to support the technological transformation of the German industry (Kagermann et al., 2013). One of the forms of support was (and still is) the "Plattform Industrie 4.0". The Platform plays a very important role in this concept because it is the centre of knowledge for stakeholders. The structure of this platform is based on multifaceted cooperation between government agencies and the private sector on issues of the fourth generation industry.

Emerging more than a decade ago by the federal government in Germany, with the support of corporations and think tanks, the concept of 'Industry 4.0' has gained many supporters. The governments of successive countries, following the German Platform, have incorporated the assumptions of this new concept into the strategic directions for the development of their economies and industries.

The idea of the concept "Industry 4.0", with the broad picture of the changes in the Fourth Industrial Revolution, was presented in the book *The Fourth Industrial Revolution* by Klaus Schwab (2016)—a president of the World Economic Forum. This book is considered one of the most important items of contemporary literature describing the issues of digital transformation. In his book, K. Schwab paints a picture of the key technologies driving the digital revolution and points out the most important consequences they will have for governments, the business world, society, and everyday life. The Fourth Industrial Revolution is different from the previous revolutions because it creates a new quality of life and a new business model.

The previous industrial revolutions streamlined manufacturing processes and enabled mass production. In the last two revolutions, Information and Communication Technologies (ICTs) have become opportunities for development. The use of PLC controllers, industrial software systems, CAD, CAM, MES, ERP, PLM, SCADA, etc., as well as the introduction of industrial robots in the production halls allowed full control of the production process in the plant. Production has become much more transparent, and the use of the so-called "flexible production systems" have resulted in previously unattainable efficiency and quality, while allowing custom-made products. The Fourth Industrial Revolution is characterized by a series of new technologies that connect the physical, digital, and biological worlds, affecting all disciplines, economies, and industries. The world has the potential to connect billions of people to digital networks, radically improve organizational performance, and even manage resources in ways that can help regenerate the environment, potentially reversing the damage done by the previous industrial revolutions.

The Fourth Industrial Revolution brings radical changes for workers, organizations, and society as a whole. By disrupting the incentives, rules, and behaviors of economic urban life, this revolution will transform how we communicate, learn, behave, and relate to each other and to our surroundings. While this era does not come without its challenges, it provides us with an opportunity and responsibility to shape how we want to live and work in our cities in the future, and allows us to unite global communities and build sustainable economies. Just like the industrial revolutions of the past, the Fourth Industrial Revolution is leading to a complete societal shift in which we can proactively shape to be inclusive and human-centric giving the movement purpose and structure (Schwab, 2016; 2018).

The principles of the Fourth Industrial Revolution have been adopted in many countries who have labeled it "smart industry", "advanced manufacturing", or the "Industrial Internet of Things", or "Industrie 4.0" (based on the documents of European Parliament, 2016 and the European Commission, 2015). In many countries, the technological changes attributed to the Fourth Industrial Revolution are called simply "Industry 4.0".

The Fourth Industrial Revolution, with the concept of Industry 4.0, represents a new paradigm in the management and organization of the manufacturing and industrial value chain. Industry 4.0 is based on the connected physical manufacturing technologies and digital technologies such as Big Data, Artificial Intelligence (AI) and Cloud Computing (Fatorachian and Kazemi, 2018; Thoben et al., 2017). Peters (2016, 2017) points out in his speeches about Industry 4.0 (case study: steel sector) that the industrial transformation is the result of the 'marriage' of the physical and digital worlds.

The Fourth Revolution is a combination of technological breakthroughs and incredible achievements in areas as vast as AI, robotics, the Internet of Things (IoT), autonomous cars, 3D printing, nanotechnology, biotechnology, materials science, energy storage, and quantum computing—to name a few. Many of these innovations will create technology that unites the physical, digital, and biological worlds (Schwab, 2017).

The Fourth Industrial Revolution is "characterized by a fusion revolution of technologies that is blurring the lines between the physical, digital, and biological spheres" (Schwab, 2016). In the digital and smart era, ICTs are increasingly being recognized as a tool for facilitating socio-economic transformation and inclusive growth (World Economic Forum, 2016; United Nations, 2016; European Commission, 2015). The Organisation for Economic Co-operation and Development (OECD) defines inclusive growth as "economic growth that creates opportunity for all segments of the population and distributes the dividends of increased prosperity, both in monetary and non-monetary terms, fairly across society" (OECD, 2016).

Changes in modern economies and industries based on the technologies of the Fourth Industrial Revolution are referred to as digital transformation, which is defined in two ways. Firstly, it is identified with typical digitization, and secondly, at the same time, digital transformation means the integration of digital technology into all areas of a company or factory. In the Third Industrial Revolution, the term "Digital Manufacturing" referred to the use of digital controls on the production line (as opposed to analogue controls), and in the Fourth Revolution it means a much broader scope, referring to the combination of manufacturing technology, networked information technology, and information analysis to provide better understanding, coordination, and control of production processes. Digital transformation involves the integration of technology and digital solutions in every area of a company's operations. It is not only a technological change but also, in a sense, a cultural one. It requires the company to significantly transform the way it operates and to take care of the experience and benefits offered to customers. Digital transformation is a fundamental change in the way we think about customer expirence, business model and process. It's about finding new ways to deliver value, generate revenue, and increase efficiency.

Thanks to digital transformation, it is possible to use data collected by industrial companies through Industrial Internet of Things (IIoT) systems to create innovative services and products. This can be data of a technical, economic, or organizational nature. Based on them, decisions can then be made to control production, processes, or orders. It also makes it possible to expand existing market offers, both manufacturing and service, so that they increase quality and customer satisfaction (Culot et al., 2020). This satisfaction is now greater because the customer can receive a personalized or 'tailor-made' product. In addition, the customer is an active participant in the product design, manufacturing, and distribution processes by communication systems, and visualization of processes on his (her) mobile devices. Digitalization is actively promoting the development of new businesses and new forms of manufacturing and customer service.

Many researchers (Castagnoli et al., 2021; Neubert, 2019; Santos-Pereira et al., 2022) agree that the digital transformation has internationalized business in the global marketplace. IC systems combine all available computer and digital technologies into a common space for data transfer and communication via the Internet, wireless networks, Bluetooth networks, smartphones, etc. Data is stored on portable memory devices, hard drives, CD/DVDs, tapes, servers, clusters, or stored in the cloud. New technologies are a collection of computing devices and solutions used to process and transfer data at a higher level of abstraction than just hardware. Feliciano-Cestero et al. (2023) documented the impact of people and elements such as knowledge, digital servicing, and leadership on the impact on internationalization at the individual, corporate, and macro levels. Naglic et al. (2020) confirmed the effectiveness of digitization in increasing export performance. Ilcus (2018) argued that digitalization creates a new competitive environment by transforming business strategies, models, and processes.

The key direction of digitization in manufacturing enterprises is processes optimization. According to Bouwman et al. (2017) and Müller (2019), digitization can reduce resource loss in business processes. New technologies, through the precision of the operations performed, as well as the great opportunity to learn from historical data, are a strong support for machine operators in the maintenance optimization (Pinciroli et al., 2023) and for operational managers in the decision-making process. Cyber-Physical Systems (CPSs) are able to effectively support the optimization of production processes by connecting different types of elements (IT, mechanical, etc.) and interaction with physical entities (Csalódi et al., 2021).

The scope of digital transformation includes all activities related to the modification of business and organizational processes, as well as production and logistics processes, including the digitization of production. The latter involves the introduction of digital technologies into production processes. The goal of digital transformation is to take full advantage of the opportunities brought by modern digital technological solutions. The Fourth Industrial Revolution is characterized by future industry development trends to achieve more intelligent manufacturing processes, including reliance and construction of CPSs, and the implementation and operation of smart industries that use advanced techniques and technologies (Schwab, 2016; Zhou, Liu and Zhou, 2016; EC, 2015; Gajdzik, 2020; Gajdzik, et al., 2019).

The modern digital transfromation is a key component of business strategies, and while it is not the only component, it is fundamental to the success or failure of the effort. The right technologies, as well as people, processes and operations, enable organizations to respond quickly to disruptions and opportunities and meet new and changing customer needs. They prepare organizations for future growth and foster innovation, often in unexpected ways. Moreover, digital transformation, especially in manufacturing, involves the use of Industry 4.0, IoT, and AI solutions (Kurfess et al., 2020).

Business process transformation is clearly impacting business operations as a result of improvements in workflow management. For example, implementing a cloud-based digital supply chain management system reduces downtime, improves production, and increases profitability. In the automotive industry, digital technologies enable centralization and automation of business models and billing processes, including forward-looking subscription-based processes similar to what is currently happening with software (Deloitte, 2020). In the traditional industry (continuous processes), including metallurgy, digital technologies enable steelmakers to work more closely with their customer markets, including the automotive sector, which has increasing performance requirements for steel products (Gajdzik, 2022a). Today's ubiquitous pull systems initiate manufacturing processes, hence the need to tighten relationships with customers and plan production under the customer.

Technological trends in the concept of Industry 4.0 and the digital transformation happening can be divided into several levels. The base level is low-application modern digital innovations. The higher level is high application level innovations or advanced manufacturing technologies that enable the digital industrial revolution (Ciffolilli and Muscio, 2018; Kumar et al., 2020). Smart sensors, industrial robots, smart wearables, and machine controllers are examples of low-level digital transformation technology trends that can be acquired and implemented as discrete digitization projects in industrial environments (Frank et al., 2019). The higher technology trends of Industry 4.0, such as the Industrial Internet of Things (IIoT), Cyber-Physical Production Systems (CPPS), or digital twins, are based on the integration of various low-level digital and operational technologies such as network infrastructure, sensors, machines, and even connected human components (Ciffolilli and Muscio, 2018; Boyes et al., 2018; Kumar et al., 2020; Drath and Horch, 2014; Ghobakhloo and Iranmanesh, 2021). The highest level of the digital transformation, to date, is called ubiquitous integration, or horizontal integration and vertical integration through the capabilities of technology to deliver real-time data and customer orientation (Dev et al., 2020).

Rapid adoption of new and emerging technologies such as the IoT and Internet of Services (IoS) have given rise to the Fourth Industrial Revolution. Most the literature focuses on the technological aspect. However, the implementation of most emerging technologies in society and its promotion has revealed more significant complexity, including various non-technological elements, such as social, legal, institutional. This multidimensional complexity requires an appropriate legal ram. Innovations related to digital transformation, such as smart cities, smart mobility, and smart industries, require governments to implement regulatory mechanisms to support their successful adoption. Governmental policies and legislation play an important role in managing a complex digital system, and a connected and smart environment (Scholl and Al-Awadhi, 2016; Manda and Backhouse, 2017).

The Fourth Industrial Revolution brings with it changes that are challenging, in the areas of security, trust, accountability, and privacy of personal data. Ongoing changes require governmental regulation (Zhou, Liu, and Zhou, 2015). As Saniuk et al. (2020) and Grabowska et al. (2022) point out, the dilemmas of the uncontrolled development of digital manufacturing may concern, for example, intellectual property rights in the case of co-creation of products by customers, the impact of personalized products on the environment, corporate responsibility for the actions of AI and others. One of the challenges is security and privacy. Security and data privacy issues have arguably become one of the most significant concerns in the Fourth Industrial Revolution, where technology has become a driver (Waidner and Kasper, 2016). Integration of systems in the Fourth Industrial Revolution requires the development of new security and protection mechanisms for the faster and more flexible collaborative value networks, and smart production systems (Waidner and Kasper, 2016). Moreover, modern business needs infrastructure. Developing countries are not only confronted by societal challenges but technological and infrastructure challenges. Poor technological infrastructure and communication networks are barriers for business development in the Fourth Industrial Revolution. Broadband penetration, for example, is still low in developing countries compared to developed economies that are considered leaders in broadband and other ICT infrastructure (International Telecommunication Union, 2023).

With the rapid implementation of digital technologies in business processes, the demand for a highly skilled workforce is growing rapidly. During the World Economic Forum 2016, participants predicted an increase in demand for IT staff. Digital transformation and innovation, therefore, require a new race of workers and citizens, one that is skilled, innovative, and technologically savvy. Governments are now focused on developing the skills of the future that create digital skills, technical skills, and soft skills (interpersonal) (EC: e-Skills for growth and jobs). People learn to work with new technologies that help them in many areas, but which require human interaction (Loretntz et al., 2015; Romero et al., 2016). Work and the importance of work are changing for workers, and organizations must now balance the effectiveness of technological innovation with new jobs and employment concepts, as well as global and local shifts of power. Latest research in the context of the Fourth Industrial Revolution highlights that jobs will be, more than ever, driven by digitalization and robust smart ecosystems (Chearavanont, 2020). Therefore, AI and machine learning will increase in importance and provide the foundation for which people conduct their work and business. This means that companies and employees will have to re-imagine what new ways of working might look like and how they will fit into new workplace designs and concepts (Ross and Maynard, 2021).

The process of digital transformation should be aligned with the culture and values of the organization or the company. A loss of confidence in the corporate culture within a company can have a negative impact on productivity, initiative and staff well-being. Slow or pessimistic adoption of new digital technologies can lead to missed goals and loss of competitiveness, revenue, and brand value (Ustundag and Cevikcan, 2018). Moreover, it is assumed that the high technologies of the Fourth Industrial Revolution increase the resilience of business to the instability of the environment stated that digital technologies increase capacity responsiveness and capacity flexibility, that could help predict and mitigate disruption by enabling better visibility, forecasting accuracy, flexibility, and coordination in real-time environments throughout the supply chain (Ivanov and Dolgui, 2019).

In summary, the Fourth Industrial Revolution is a significant event for the development of industries, economies, and society. There are many social and economic and business trends in the ongoing digital transformation (Gajdzik et al., 2021b). Promoted for more than a decade, the concept of Industry 4.0 is a megatrend that is having a strong impact on many areas of 21st century civilization. This new concept poses challenges for governments, companies, scientists, and other groups. The need for investment, changing business models, data issues, legal liability and intellectual issues, features of personalized products, labor standards in CPSs, and skill mismatches are all significant challenges for economies and societies in the Fourth Industrial Revolution.

## 2. Framework of Industrial Transformation

Modern industrial transformation is based on the concept of Industry 4.0 and its technologies. Changes in industries are concerted on CPSs, computer systems, virtual reality, IoT, IoS, and cloud computing. The concept of modern industrial transformation is based on the automation and robotization of operations and the digitalization of the manufacturing process and logistics (Kagermann, 2014a). Automation (from the Greek 'automatos', meaning self-acting) is the process of reducing human physical as well as mental work by replacing it with machines and devices that perform repetitive tasks automatically. Automation means the use of machines and computers that can operate without needing human control.

Machines and technologies, instead of people, work in factory (Definition of *automation* from the *Cambridge Academic Content Dictionary*, Cambridge University Press). Automation in industries is identified with the industrial process, where all possible operations that are performed are transformed from a manual process to an automated or mechanized one (Groover, 2024 in *Encyclopedia*

*Britannica*). High automation involves the use of devices to collect and process data and, above all, to take over the cognitive, intellectual, and decision-making activities of humans. Technologies replace humans during the use of machinery (e.g., machine tools), vehicles (cars, airplanes), management of objects (smart home), or during creative work (e.g., designing, constructing, teaching). Although automation is commonly identified with robotization, these concepts are not the same. Robotization is the form of the introduction of robots to carry out industrial tasks (*Collins Dictionary*). Industrial and service robots are particularly important for the development of industry. According to the *IFR Report*, 3.5 million robots are working in the world (IRF, Report 2022).

Automation and robotization are integrated into modern business solutions. Digitization is a constant convergence of real and virtual worlds. In the context of the rapid development of ICTs, digitalization is the main driver of innovation and change in all sectors of the economy. Digitalization is the use of digital technologies to change a business model and provide new revenue- and value-producing opportunities; it is the process of moving to a digital business (Gartner Glossary).

In the modern digital transformation, special attention is paid to the aspect of integrating digital technologies into the enterprise ecosystem (Pollak, 2022). Digitalization networks people and things and enables the convergence of the real and virtual worlds through ICTs (Kagermann, 2014a). The Fourth Industrial Revolution refers to the further digitalization and integration of information technologies including applications such as the IoT (Lu, 2017), Cloud Computing (Lu, 2017), cobots (Bortolini et al., 2017), Big Data Analytics (Wang et al., 2016), additive manufacturing (Hofmann and Rüsch, 2017), and CPSs (Liu et al., 2017). These integrated technologies enable a "smart factory" (Frank et al., 2019; Osterrieder et al., 2020), in which humans, machines, and products communicate with each other via both physical and virtual means (Kagermann et al., 2013), and can contribute to increased sustainability (Bai et al., 2020). According to BDI (2015), the digital transformation takes effect via four levers: digital data, automation, connectivity, and digital customer access (Table 1).

According to Lasi et al. (2014), the following three components, also termed Industry 4.0 (I4.0), make up digital transformation:

- *digitalization and enhanced integration across different vertical and horizontal value chains:* creating personalized products, digital customer orders, automated data delivery systems, and integrated customer service solutions;
- *product and service digitalization:* providing comprehensive smart grid descriptions of the product and associated services;
- *implementation of innovative digital business concepts:* high interaction within systems and technological capabilities for creating new and integrated digital business models.

**Table 1** Four levers of digital transformation in manufacturing industries.

| Levers | Functions | Solutions | Effects |
|---|---|---|---|
| Digital data | Capturing, processing and analyzing digital data. | Wearables, Big Data, Internet of Things, sensors, analytical technologies. | Digital data allows better predictions and decisions to be made. |
| Automation | Combining traditional technologies with Artificial Intelligence (AI) is increasingly giving rise to systems that work autonomously and organize themselves. | Robotics, additive manufacturing, drones, autonomous vehicles. | Automation reduces error rates, adds speed and cuts operating costs. |
| Connectivity | Interconnecting the entire value chain via mobile or fixed-line high-bandwidth telecom networks. | Cloud computing, broadband, databased, digital products, Machine-to-Machine (M2M). | Communication technologies, supported by computer systems, synchronize supply chains, and shorten both production lead times and innovation cycles. |
| Digital customer access | Online sales, communication via web applications, electronic settlement of transactions, etc. | Mobile Internet/ apps., social networks, e-commerce. | The (mobile) Internet gives new intermediaries direct access to customers to whom they can offer full transparency and new kinds of services. |

*Source:* Own elaboration based on the digital transformation of industry. BDI, 2015.

The foundation of the industrial Internet involves the embedded accessibility and monitoring of systems throughout the entire enterprise in real time. IIoT platform combines the level of operating technology with the level of IT technology. Sensor data generated in production facilities can be easily read and processed and used as a basis for sustainable business decisions. According to (PwC), IIoT is a key technology in digital backbone. For IoT to be able to combine OT and IT, we need IT systems to support processes. Key systems in manufacturing companies are: ERP (Enterprise Resource Planning) and MES/MOM (Manufacturing Execution System/Manufacturing Operations Management) (Table 2). Companies that already have a fundamental IT infrastructure can invest in additional systems and technologies, including CRM, IBP, WMS, PLM, as well as maintenance, quality control, and more (PwC, 2022). Support for ERP and MES are SCADA/DCS and SPS systems and sensors. A key element of an effective hyper-automation process are also low-code solutions (Table 2).

**Table 2**  Digital backbone of industrial transformation

| *IC Systems* | *Description* |
|---|---|
| ERP | ERP (Enterprise Resource Planning) systems have made great improvements over the years. Industry 4.0 technologies have made some milestone improvements to the firms in context with the ERP systems such as data acquiring, analyzing, storing, and decision-making capabilities. Nowadays, information technology has become an essential tool for the operation and management of all activities ranging from production scheduling to supply chain management. The systems integrate the basic business functions such as production, finance, and marketing as well as side functions such as cost accounting, purchasing, distribution, customer relations, cash flows, warehouse management, human resources, material management, electronic banking, and quality control (Akyurt et al., 2021). |
| MES | MES (Manufacturing Execution System) is a system for managing and controlling the production area, which constantly monitors the efficiency of the manufacturing process. The system already used in production to track and manage production operations. The system helps to monitor production progress, collect data from machines and operations, optimize production processes, and manage human and material resources. Main MES blocks: Human Machine Interface (HMI), Data Acquisition Systems, Quality Management Systems, Production Planning and Scheduling. |
| MES+ERP | ERP (Enterprise Resource Planning) system with MES (Manufacturing Resource Planning) cover almost the entire product cycle from production to sale. This gives incredible possibilities to control and maintain constant quality and condition of manufactured products. |
| PLM | PLM (Product Life Cycle Management Solutions) is a software that manages all information and processes at every stage of the product or service lifecycle in globalized supply chains. This system includes data from items, parts, products, documents, requirements, engineering change orders, and quality workflows. |
| IIoT | IIoT (Industrial Internet of Things) is a platform that combines the level of operational technology with the level of IT technology. Sensor data generated in production facilities can be easily read and processed and used as a basis for sustainable business decisions. |
| Low Code Automation | Low-code solutions enable business and IT teams to collaborate effectively in the development of applications by providing a common platform for work, prototyping capabilities and rapid validation of created solutions. This allows you to create products that are perfectly adapted to market needs, while at the same time being fully flexible and scalable. |

*Source:* Own elaboration based on Digital Factory Transformation Survey 2022, PwC.

The researcher (Dubey et al., 2022) mention the following components of a transforming enterprise: ERP, IoT, IIoT, Big Data, Man-made consciousness (AI), M2M, Digital Factory, Smart Manufacturing, AI, Distributed computing, Real-time information preparing, Integration computer systems (ERP, MES, BI, SCADA), and (CPSs). The automation pyramid of a manufacturing company includes: ERP system containing business logic and production management modules MES (Manufacturing Execution System), HMI operator panel, and barcode scanner.

The ERP system and its current structure is the result of the evolution of various smaller systems to support different departments or organizational functions, such as inventory control or material requirements planning (Law, 2019: 1–3). Nowadays, managing business processes using computer systems means having data from all departments in one database and access to information reflecting the current state to the second (Wrycza and Maślankowski, 2019: 414). The ERP systems are linked to MES systems and together they create a complete picture of the process flow and open up possibilities for their optimization.

The next step is the process of acquisition of production data powering the IT and computer systems. For this purpose, an HMI (Human Machine Interface) operator panel is created and barcode scanners are used. An HMI is an industrial interface between a machine or process (computer program or system) and the human operator—the operator (McMillan and Vegas, 2019: 472).

The automation pyramid grows and expands at hierarchical levels as the company absorbs new technologies. The basis of process digitization is full automation, such as purchasing components, cabling, and investing in devices that generate production data and interact with cloud computing. Cloud services allow companies to remotely access, control and manage machines, processes, and data in real time from any location, allowing you to collect, process, analyze, and share data on an ongoing basis. Cloud computing gives users the flexibility to access data and allows them to choose the software provider. Manufacturing companies use the analytical capabilities of cloud computing to handle large volumes of data generated, among others, by mounted sensors (Grossman et al., 2009; Jain et al., 2017; Tiwari et al., 2013; Tiwari, 2021). Modern digital technologies determine the direction of data transfer, from the source of its creation to the next levels in the enterprise hierarchy.

According to Ghobakhloo and Iranmanesh (2021) the determinants of digital transformation success are: technological factors, organizational factors, and environmental factors. Technological factors are conditioned by technological trends (IIoT, AI, AVR, CPS etc.), they create design principles of Industry 4.0 such as value chain integration, interoperability, real-time capability, additive manufacturing, visulaization, etc. According to Roblek et al. (2016) the transformation of enterprises to Industry 4.0 requires the following basic changes:

- digitalization of production at the level of management and production planning,
- automation—systems for acquiring data from production lines and from individual machines,
- linking production plants within a comprehensive supply chain—automatic data exchange.

These components are integrated into CPSs, which are the connection between the real and virtual worlds. Machine communication (M2M) and intelligent products are integrated into CPSs. In the process, the IoT is the enabler of M2M communication (Greengard, 2015; Kagermann, 2014b). Technologies of Industry 4.0 are opportunities for improved productivity, innovation, customer satisfaction, business model innovation, enhanced flexibility, resources efficiency, modularity, service orientation, etc.

Ghobakhloo and Iranmanesh (2021) assume that the digital transformation under Industry 4.0 is a strategic business change that relies on the institutionalization and integration of various combinations of modern Information and Digital Technologies (IDTs) including AI, data analytics, digital twins, industrial robots, and blockchain technology. This process also involves adopting agility, customer orientation, and product individualization as core competencies (Fatorachian and Kazemi, 2018; Kagermann, 2015; Yaqub and Alsabban, 2023). They state that the success of the digital transformation in the Fourth Industrial Revolution requires leading Industry 4.0 technologies, which are: blockchain technology, Big Data Analytics, Cloud Computing, AI, and IoT (Table 3).

Industrial transformation assumes that collaborative high technologies will reach a higher level of development, which is typified by AI. Smart and collaborative technologies are already setting the stage for the development of industry into smart factories and making it part of the new market economy. According to Kagermann (2014a), digitization is the most powerful driver of innovation and a strong incentive for innovation. Kagermann predicted that digitization of processes will transform all key industry sectors, such as manufacturing, energy, mobility, healthcare (Kagermann, 2014a). The exponentially increasing amount of data (Big Data) and the convergence of various affordable technologies that came with the eventual establishment of ICTs are already transforming key areas of the economy, and with regard to industry, are creating new dimensions of operation referred to as the smart factory. Smart factories, that are the unification of software and hardware devices, constitute the main framework of Industry 4.0. Smart factories are the manufacturing environments where workers, machines, and resources communicate and cooperate in complex manufacturing networks. The integration of AI into the manufacturing systems will facilitate to set up networks, and enable them to learn, infer, and act autonomously based on the collected data during the industrial process (Dopico et al., 2016). AI focuses on computer programs that are capable of making their own decision to solve a problem of interest, and the systems that are created with AI are intended to mimic the intelligent behavior of expertise (Kumar, 2017). The image of a smart factory appears as a system of CPSs controlling physical processes and creating virtual (digital) copies of the real world, and making decentralized decisions. Smart factories are dynamic, self-organizing, and optimizing technical and organizational systems (Siemens, 2017).

In smart factories, all technologies collaborate with each other through the IoT, which can connect people, products, and machines into one coherent

**Table 3** Leading technologies of digital transformation in Industry 4.0.

| Technology | Description | Opportunities |
|---|---|---|
| Blockchain technology | Blockchain technology uses established encryption rules and serves as a repository for transaction information, and is documented and shared through a decentralized peer-to-peer network. All participants maintain a copy of the digital blockchain ledger and verify new entries using a consensus protocol on this network (Yaqub and Alsabban, 2023). | The use of blockchain expands SCM and increases its operational efficiency. BCT provides maximum visibility throughout the supply chain, with product status verification, location tracking, and other useful features contributing to supply chain efficiency, flexibility, agility, and collaboration, which, when combined with other I 4.0 technologies such as IoT and Big Data Analytics, can help realize diversification and achieve great mass-scale personalization In addition, the technology can be used for processes such as real-time information exchange, cybersecurity, transparency, reliability, traceability, and visibility. Blockchain technology focuses on the exchange of value. The technology is particularly important for supply chain networks to make and verify transactions for organizations and individuals without a central control authority (Yaqub and Alsabban, 2023; Aslam et al., 2021; Lezzi et al., 2018; Banerjee, 2019). |
| Big Data Analytics | Big Data is a technological trend that helps to better evaluate or explore the world and consequently streamline decision-making, communication, coordination, and action (AlBar and Hoque, 2017; Aurelie, 2008; Suoniemi et al., 2020; Xu et al., 2019). Big data is a huge amount of data generated by sensors and products. | Big Data comes from public or private organizations (Zhan et al., 2018). Many companies have invested in collecting, integrating, analyzing, and using data to tailor their operations to meet customer needs. Recent advances in Big Data Analytics have resulted in methods to increase the value for each consumer (Yaqub and Alsabban, 2023). Big Data is being applied to real-time process optimization as well as lean manufacturing operations, and its use has implications for business model innovation and B2B relationships (Yaqub and Alsabban, 2023; Dai and Liang, 2022; Hallikainen et al., 2020; Gupta et al., 2020; Sodhi, 2020). |

*Contd.*

**Table 3**  *Contd.*

| Technology | Description | Opportunities |
|---|---|---|
| Cloud Computing | Cloud Computing refers to a computing paradigm that allocates tasks across a set of connected devices, applications, and shared services that are accessible through a network. Commonly, this interconnected network of servers and links is referred to as "the cloud" (Yaqub and Alsabban, 2023). The cloud is made up of multiple entities including clients, computing servers, and shared hosts (Grossman et al., 2009, Jani et al., 2017, Tiwari and Jani, 2013; Tiwari, 2021). | Cloud Computing, together with its closely aligned IoT data management tools such as fog computing and edge computing, through enabling real-time data exchange through cloud, fog, and/or edge (data layers)-based platforms, has reshaped the way global enterprise networks connect and work together, creating an agile, more collaborative environment within (focal or cooperative) organizations (Abdollahzadegan et al., 2013; Bonomi et al., 2012). Cloud-based manufacturing is an advanced manufacturing process in which cloud computing, IoT, and virtualization is incorporated. These technologies transform the manufacturing process such that that the services involved can be shared and circulated, thereby improving the production process to drive organizations to success (Weyer et al., 2015; Zhong et al., 2017). The adoption of cloud computing facilitates a networked supply chain through enhancing real-time visibility, which makes supply chains more dynamic, safe, and collaborative (Carvalho et al., 2012; Yaqub and Alsabban, 2023; Tiwari, 2013; Fisher et al., 2018). |
| Artificial Intelligence (AI) | AI was established to design and build "thinking machines", including a range of enabling computing technologies that have been designed to recognize, learn, think, and act in an appropriate manner (Yaqub and Alsabban, 2023). | When AI comes into play, the manufacturing process has reasoning, learning, and acting possibilities, which contribute significantly to intelligent production (Dash et al., 2019; Kumar et al, 2019; Eliasy and Przychodzen, 2020). Moreover, the use of AI minimizes the number of human errors, enables robotic training, handles inventory management, increases precision, and ensures maximum responsiveness and adaptability to the external environment (Chowdhury et al., 2022). AI aids in handling consumer needs and aims to achieve the mass customization of products by smart design (Saldivar et al., 2016). Intelligent manufacturing, a self-regulatory and self-controlled process, facilitates the productions of goods according to the specified design requirements enabling differentiation and enhanced customization (Kumar et al., 2019). |

*Contd.*

**Table 3** *Contd.*

| Technology | Description | Opportunities |
|---|---|---|
| Internet of Things (IoT) | IoT corresponds to a particular technological approach where several appliances that can turn the internet on and off to use software and automation procedures that enable smart operations are networked with each other (Quetti et al., 2012). Every device that supports smart sensor technology qualifies as an IoT environment device. Every object, including humans, can become embedded in the IoT network, through smartphones or wearables (Quetti et al., 2012; Zhong et al., 2013). These smart devices are interconnected and interact with each other digitally using sensors. The devices can interact automatically and adapt to carrying out production logics (Yaqub and Alsabban, 2023). | IoT represents an innovation that has transformed how we arrange and organize our work and home lives, how we navigate our ways around and use transportation, and how we control industrial machinery (Leiting et al., 2022; Waschull et al., 2020). Cui et al. (2020) described IoT as a pathway towards achieving operational excellence.<br>Thanks to IoT, improvements in the supply chain are constantly possible, which become agile chains (Lou et al., 2011). Having IoT capability can enable a company to make full and comprehensive use of AI, simulation, automation, robotics, data acquisition systems, and networks for advanced engineering. With these advanced solutions, a powerful, collaborative, and future-proof manufacturing infrastructure can be built that can deliver sustainable cost benefits (Yaqub and Alsabban, 2023), Moreover, information on customers' purchasing and usage patterns gathered through sensors on products integrated into IoT may lead to a better and comprehensive understanding of customers' characteristics and their tastes, preferences, and requirements (Benias and Markopoulos, 2017; Nagy et al., 2018; Weng, 2020, Ng et al., 2015; Yaqub and Alsabban, 2023). |

*Source:* Own elaboration based the literature indicated in the table.

mechanism, delivering new quality products and services. It is assumed that as technology evolves, products, means of transport, and tools will 'negotiate' with data, automation, and broadband networks (Aheleroff et al., 2020). Technological communication and collaboration will be focused on finding the most effective ways of functioning value chains to deliver the expected and designed product. Product personalization is a challenge for high-tech and ubiquitous machine communication (M2M) and interaction machines with people. The term for personalization is 'tailor-made' products (Shohin and Runyang, 2018).

The industrial transformation taking place is an inversion of the logic of the conventional production process, it means that the industrial production machine no longer simply 'processes' the product, but that the product communicates with the machine to tell it exactly what to do. The technologies are autonomous, self-monitoring, and adaptive to different situations (Sniderman et al., 2016; Ślusarczyk, 2018).

Many scientists agree that the concept of Industry 4.0 needs to digitize and automate the manufacturing environment, as well as create digital value chains to enable communication between products, environments, and business partners (Lasi et al., 2014: 239–242; Hermann et al., 2016: 3928–3937; Lu, 2017: 1–10; Kagermann, 2014a). Future industry will be based on multiple high technologies and agile value chains (Hermann, 2016: 3928–3937). For Lu (2017: 1–10), the industry concept termed Industry 4.0 is a form of integrated, customized, optimized, service-oriented, and interoperable manufacturing process that is correlated with algorithms, big data, and high technology.

Substitutes for the name "Industry 4.0" are "Smart Manufacturing", "IoT", "Smart Factory" (Sinderman et al., 2016). Germany uses the Industrie 4.0, US – Industrial Internet, China – Made in CHINA 2025 and Internet Plus, Taiwan – Productivity 4.0, Indie – Industry 4.0: Make in India and Skill Indie, Japanese – Industrial Value Chain Initiative, France – Nouvelle France Industrielle, Netherlands – Smart Industry, UK – High Value Manufacturing Catapult (HMV Catapult), Spain – Industria Conectada 4.0, Poland – Future Industry.

The concept of Industry 4.0 is described by the key technologies of the Fourth Industrial Revolution, which are called the pillars of the concept. Most often, which are reflected in publications (Erboz, 2017; Burrell, 2019; Senn, 2019; Boston Consulting Group, 2015; Dubey et al., 2022), and identified as the nine pillars of the Industry 4.0 concept (Table 4).

One may wonder if each of the nine pillars of Industry 4.0 is equally important to the development of a smart factory. In the Siemens *Guide for Managers and Engineers* (2017), we find the grouped technologies, or pillars of Industry 4.0, into two categories. The first is made up of technologies that are called the "core of the Smart Factory", and the second is made up of new and popularizing industrial technologies. Details of the grouping are included in Tables 5 and 6.

**Table 4** Nine pillars of Industry 4.0.

| No. | Pillars | Description |
|---|---|---|
| 1. | Big Data and Analytics | Technologies used to collect and analyze large amounts of data from a variety of sources to support corporate decision making. Historical and current data form huge collections that are supported by software with algorithms based on AI and Machine Learning (ML). Thanks to these collections and solutions, industrial technologies are becoming increasingly autonomous. |
| 2. | Internet of Things (IoT) | Connecting technology and sensors to a network that enables objects, machines, and finished products to communicate and interact with each other and with people. Decision-making systems are decentralized and act in real time through access to historical and current data. |
| 3. | Horizontal and Vertical Integration | Systems' integration takes place both within the company and throughout the value chain (all departments and functions of the company are part of one comprehensive system). Vertical and horizontal integration creates a complete picture of an organization, which is a complex system consisting of other components. The integration of information systems is the basis for creating a complete and integrated silos-free mold from the factory. |
| 4. | Cloud Computing | With access to more and more data and the development of ICTs, cloud computing technologies are developing. Public, private, and hybrid clouds are available on the market. Clouds are a collection of business applications and services, including process control applications and software services. |
| 5. | Additive Manufacturing | Additive manufacturing is most often associated with 3D printers. 3D printing is currently used for prototyping, for the production of special components, spare parts, and highly personalized batches of products. Companies that have 3D printers produce products for their needs (such as printing spare parts for machines) or commercial products. |
| 6. | Augmented Reality | It includes systems that add multimedia information to people's perceived reality by using mobile, visual, auditory, or manipulative devices. |
| 7. | Autonomous Robots and Artificial Intelligence | A new generation of industrial robots with more functions than previous ones that can cooperate with humans by interacting with them. In addition, the latest generation of robots are able to communicate with each other, and learn from historical data. |
| 8. | Computer Models and Simulation | The models are used to test and optimize machines, to test new products and processes, and to predict problems before they occur. One such simulation model is the digital twin, a virtual replica of a physical object in which planned solutions can be mapped and design errors can be detected before they move to the implementation phase. |
| 9. | Cybersecurity | Technologies to protect factory production systems and IT networks against potential threats using cybersecurity programs. |

*Source:* Own elaboration based on Rüßmann et al. (2015); Pollak (2022).

**Table 5**   Core technologies in smart factories.

| Technology | Features and Functions |
|---|---|
| Industrial Internet of Things (IoT) | • implementation of technical and business solutions based on IT technologies,<br>• communication with distributed sensors and network devices. |
| Data Analytics and Production Optimization | • use of real-time data processing and analysis software,<br>• availability of current production information at the management level of the enterprise (management cockpits),<br>• opportunities for production optimization and predictive maintenance. |
| Integration of IT/OT (Information Technology/ Operational Technology) and Cyber-Physical Systems (CPSs). | • creation of CPS, combining mechatronic, electronic and communication systems, and software,<br>• integration of production systems with IT and business (management) levels. |
| Cybersecurity | • implementing security measures to minimize external and internal cyber threats,<br>• Safe Industrial Systems Design Strategy. |

*Source:* Own elaboration based on Siemens (2017).

**Table 6**   New and popularizing technologies of industrial transformation.

| Technology | Features and Functions |
|---|---|
| Artificial Intelligence (AI) | • technologies enabling machines to learn and solve complex problems,<br>• advanced decision-making algorithms (autonomous decision-making and process optimization systems),<br>• learning systems and M2M environment. |
| Additive Manufacturing - 3D Printing | • accurate and cost-effective printing of products for the company's own or commercial use,<br>• rapid prototyping of components and production of parts with unusual shapes and functions,<br>• low and medium volume production of plastics, resins, and metals. |
| Digital Twin and Digital Factory | • software for creating virtual representations of physical systems and simulating them,<br>• end-to-end Product Lifecycle Management (LCA). |
| Cloud Computing | • distributed computing structures enabling remote storage and processing of data,<br>• virtualization of resources and the ability to easily scale systems,<br>• cloud services. |
| Big Data Analytics | • analysis of large and diverse datasets using advanced analytics and AI algorithms,<br>• solutions (a broad set of techniques and tools) for interpreting production data,<br>• data used for process optimization, detection of process anomalies. |

*Contd.*

**Table 6** *Contd.*

| Technology | Features and Functions |
|---|---|
| Virtual Reality/ Augmented Reality | • support of engineers and technicians during design and maintenance work, thanks to the use of goggles or other virtual devices and augmented reality, <br> • training in virtual space for example training showing accident anomalies in work. |
| Collaborative Robots and Automated Guided Vehicles (AGV) | • robots cooperating with each other and interacting with humans, <br> • autonomous vehicles for intralogistics applications, <br> • application flexibility: easy reconfiguration and programming. |
| RFID Radio-frequency Identification | • data storage and communication with production and warehouse management systems, <br> • creating intelligent products that communicate directly with production machines. |
| Mobile Interfaces | • portable equipment providing quick access to process information, including production, logistics, maintenance, etc. |
| Blockchain | • technology for distributed registries storing transaction information, <br> • the possibility of concluding the so-called "smart contracts" between entities without the existence of a third party or institution guarantor. |
| Geolocation | • geolocation, typically using GPS or IP address, <br> • the use of geolocation technologies in logistics and management of distributed assets, vehicles, and remote teams of employees. |

*Source:* Based on Siemens (2017).

After the pandemic COVID-19, the concept Industry 4.0 was complemented by the pillars of next concept called "Industry 5.0". The new concept, Industry 5.0, takes into account lessons learned from the pandemic and the need to design an industrial system that is inherently more resilient to future shocks and stresses, while integrating the social and environmental principles of the European Green Deal. In January 2021, the European Commission published a document entitled: *Industry 5.0. Towards a Sustainable, Human-centric and Resilient European Industry*. The document highlights the values of industrial development, which include a human-centric, sustainable, and resilient industry—three pillars of Industry 5.0 (Table 7). These values were identified as stabilizing factors for industrial development in a dynamic environment in which new sources of uncertainty and risk have emerged in the operation of businesses. The last three years have seen a profound revision of the principals associated with the need to deal effectively with potential challenges to industrial development (Gajdzik, 2022d). The fundamental difference between Industry 4.0 and Industry 5.0 is that in Industry 4.0 technologies are exposed, while in Industry 5.0 the focus is on business, economic, social, and environmental values (Xu et al., 2021). According to Matthews (2018), "Industry 5.0 takes the automated and efficient concepts of Internet of Things and big data and injects them with a traditional, personalized human touch." Industry 5.0 is defined as an evolution that aims to harness the creativity of human experts working together with efficient, intelligent, and precise machines (Maddikunta et al., 2021). Nahavandi (2019) points out that Industry 5.0 brings the human workforce back into the factory, where man and machine are paired to make processes more efficient, harnessing human brainpower and creativity by integrating workflows with intelligent systems (Table 8).

**Table 7**    Industry 5.0 (I5.0) directions.

| *Pillars* | *Description* |
|---|---|
| Human-centric | One of the pillars of I5.0 is the collaboration between machines and people to improve the productivity, safety, and quality of industrial enterprises. Intelligent machines are able to collaborate with people and help them make decisions and empower them at work (Nahavandi, 2019).<br>Robots will take over more and more routine tasks, including dangerous ones for humans, while leaving humans with duties that require the use of reason and human creativity (Pizoń and Gola, 2023). |
| Sustainable | In next-generation technologies, the development of systems based on renewable energy sources and the decarbonization of industry should be more strongly targeted. In addition, the recycling economy will become more important in business strategies due to the organic nature of natural resources. However, taking sustainability fully into account in a company's strategy means much more than before. Rather than limiting the negative impact of a company, truly sustainable companies focus on increasing their positive impact. In their book, Polman and Winston call this way of doing business "Net Positive" (2021). In the same vein, John Elkington (2020) speaks of "Green Swans", referring to the company's positive disruption to make our world a better place. In other words, the strategy in Industry 5.0 means that companies become part of the solution, not part of the problem. |
| Resilient | Business will be resilient through agile and flexible technologies. Technology agility and flexibility should lead to greater resilience. A key goal of the strategy will be to create organizations that are "anti-fragile," meaning that they are able to anticipate, respond, and learn in a timely and systematic manner from any crisis, thereby ensuring stable and sustainable performance (Taleb, 2008). Resilience applies to both organizations and supply chains (Jafari et al., 2022). |

*Source:* Based on European documents: *Industry 5.0: Towards a Sustainable, Human-centric, and Resilient European Industry. Report.* European Commission (Directorate-General for Research and Innovation (European Commission) 4 January, 2021, esearch-and-innovation.ec.europa.eu/knowledge-publications-tools-and-data/publications/all-publications/industry-50-towards-sustainable-human-centric-and-resilient-european-industry_en; *Industry 5.0. Towards a Sustainable, Human-centric, and Resilient European Industry.* (R&I PAPER SERIES POLICY BRIEF – Research and Innovation; European Commission Directorate – General for Research and Innovation Directorate F, Prosperity Unit F.5, Industry 5.0), and literature review.

**Table 8**   Industry 4.0 vs. Industry 5.0.

| *Industry 4.0* | *Industry 5.0* |
| --- | --- |
| • focus on increased efficiency – thanks to digital connectivity and artificial intelligence;<br>• technology – focused on the emergence of cyber-physical systems;<br>• adapting business models to optimize within existing capital market dynamics and economic models (i.e., cost minimization, profit maximization, production, and operational efficiency);<br>• lack of focus on the design dimensions and actions needed for system transformation;<br>• failure to decouple the use of resources and materials from the negative impacts on the environment, climate, and society. | • providing a framework for industry that combines competitiveness and sustainability, enabling industry to fully exploit its potential as one of the pillars of transformation;<br>• highlighting the impact of alternative management (technology) on sustainability and resilience;<br>• empowering workers through the use of digital devices (supporting a people-centred approach to technology);<br>• building transition paths towards environmentally sustainable applications of technologies;<br>• extending corporate responsibility across value chains;<br>• the introduction of indicators (for each industrial ecosystem) that show progress towards prosperity, resilience, and overall sustainability. |

*Source:* Przemysł Przyszłości (gov.pl) based on *Industry 5.0: A Transformative Vision for Europe. Governing Systemic Transformations towards a Sustainable Industry.* ESIR Policy Brief No. 3. (European Commission, Directorate – General for Research and Innovation, Directorate G – Common Policy Centre, Unit G1 – Common R&I Strategy & Foresight Service), 2021.

Although manufacturing companies are currently at a transition point between Industry 4.0 and Industry 5.0, a new age of industry is already emerging. Factories in the near future will be more strongly oriented toward people, values, and ethics. Using the nomenclature of Longo et al. (2020), the near future is the "Age of Amplification" of the role of technology vis-à-vis humans and complex social, economic, and environmental problems. Industry 5.0 will emerge when man and machine reconcile and work in perfect symbiosis with each other. Industry 5.0 is an upgrade for Industry 4.0 (Figure 1). The two concepts complement and integrate each other. Combining the two concepts makes sense because Industry 4.0 technologies are the basis for the development of Industry 5.0, which points to three key strategic directions for industrial transformation: human, sustainability, and resilience (Gajdzik, 2023a, b).

## 3.   Digitalization of Manufacturing Industries

Although Industry 4.0 is becoming increasingly popular in the industrial markets, companies (enterprises) are still asking about ways—roadmaps—to develop and implement new technologies of the Fourth Industrial Revolution. Many producers are now facing the need for digital transformation. How to do it best, what obstacles

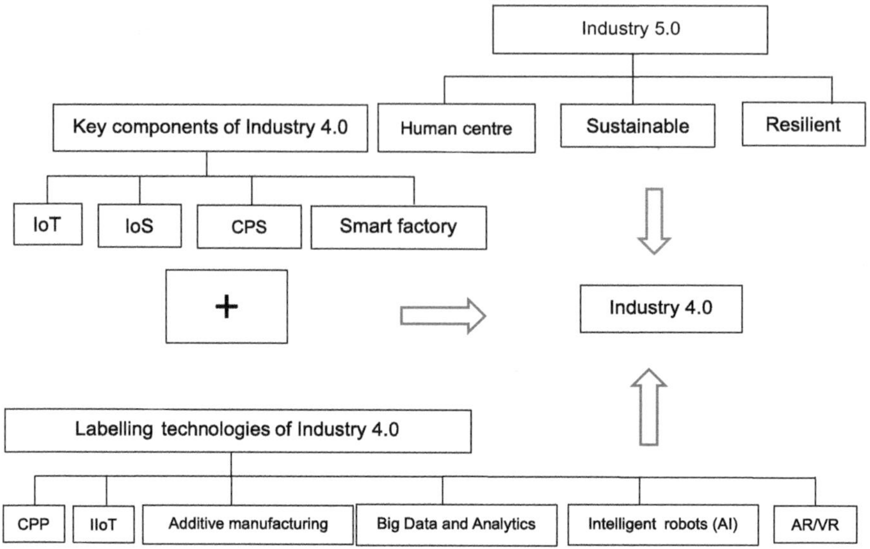


**Fig. 1** Industry 4.0 vs. Industry 5.0.
*Source:* Gajdzik, 2023b.

(barriers) can be on the way, how to overcome them? The roadmap for smart manufacturing is just a map, and each company must choose its path to smart manufacturing. Management decisions about investing in Industry 4.0 technologies are made on the basis of many analyses. The implementation of the idea of Industry 4.0 in the company focuses on projects that combine physical objects with digital solutions. This is not one project but many, spread over years or continuous. Companies interested in advanced production support technologies very often, at the diagnostic stage, use the services of the IT companies, whose engineers participate in the projects of building smart manufacturing in the company, taking into account the capabilities of the company and the specifics of the industry in which it operates (Gajdzik, 2022a).

Decisions by companies to implement the strategy simply called "the company's path to Industry 4.0" or "smart manufacturing in the company" are preceded by state analyses. In preliminary research, on identifying the market for computer-based manufacturing support technologies, industrialists are aided by the expertise of IT companies and public support institutions. Enterprises order IT teams to design technology solutions for them that will create smart manufacturing systems in the future. The IT market's service application offerings range from partial to full production automation, from intelligent sensors installed on individual machines to autonomous systems, from production-line mapping and visualization to a complete enterprise view, from a single ML algorithm to multiple process self-optimization algorithms, from individual robots to intelligent manufacturing systems. IT companies are designing solutions for individual processes and management levels for companies in a wide range of industries.

Building smart manufacturing in enterprises is step-by-step process, in which the companies must take into account the specifics of the industry to which the company belongs. The available technological solutions (offers of IT companies) consist of modules that are repeatable, but must be adapted to the specifics of the company's activities. Implementation of innovations starts with a few projects for selected scopes of activity, followed by others, and finally full digital manufacturing system architectures attributed to smart factories (Gajdzik, 2022).

Investment projects to bring enterprises closer to smart manufacturing are carried out in business areas selected by decision-makers. Investment projects are implemented in one or more business segments, and their scope includes one process or several processes at different levels of operations, such as manufacturing, service, administration, warehousing, distribution. Analyzing company reports: Deloitte (2018), PwC (2016), McKinsey (2015), Astor (2016), BCG (2015), Rüßmann et al. from BCG (2015), confirmation is obtained about the phasing (several steps) and segmentation (islands) of technology solution implementation in various industries. The implementation of projects is spread over many years, and the knowledge and experience gained from the first implemented projects allows companies to expand into further areas of change.

The scale of change (the scope of investment) depends on the already achieved degree of digitization of processes in the company, e.g., the use of ERP and MES systems. In the book *The Digital Transformation of Industry* (BDI, 2015) the three steps of digital transformation were considered as the basis of the master plan. For each stage, questions were developed to help companies determine their degree of transformation (Table 9).

The extent of the changes toward smart manufacturing varies by industries. There are industries in the market where technological progress is realized faster and easier, and there are industries that find it more difficult to make radical technological changes due to high costs. Industries that more easily adopt digital and smart technologies include the food industry, the apparel industry, home appliance manufacturers, the automotive producers, machinery industry, and the electronics industry. Slightly slower transformation to smart manufacturing is taking place in the steel industry, mining industry (mining), fuel industry, energy industry, chemical industry. Taking into account the type of production (form of production organization), it can be assumed that discrete processes are more susceptible to change towards smart manufacturing than continuous processes (Danel et al., 2023).

The effects (results) of the changes introduced in companies are also different. In the production of electricity, chemicals, or fuels, advanced technology provides greater control over processes than discrete or hybrid production, such as car production or food production. In the latter case, the changes are aimed at better adapting products to the individual needs of customers and shortening delivery times. In industries where the technology lifecycles count in decades (steel, mining, energy), the pace of change is slower than in the consumer market such as clothing, shoes, cars, household appliances. The steel, mining, fuel, and heating

**Table 9**  Three steps for digital transformation in industries.

| Steps | Questions | Outcomes |
|---|---|---|
| Step 1: Analysis of the influence of digital technologies on industry | What different future scenarios are conceivable? At what links is the value chain changing? From where are the new, scalable, platforms emerging? What are the key technologies? Which market players are affected (suppliers, competitors, customers)? Who is leading (benchmarking) the change? | Coming changes identified. |
| Step 2: Analysis of digital maturity: Where is the company at? | Where are the new opportunities for and risks to our business emerging? Which of our products, customers, and regions are affected? What (digital) capabilities can we draw on (personnel, partnerships)? Where is our digital business strategy anchored within the organization? Which products, process, and infrastructures are at risk from cyber-attacks? What is the state of our technology? | Gaps in implementation and capabilities identified. |
| Step 3: Development of an implementation roadmap | For which future scenarios do we need viable options now? What skills do we need to build up (data-processing, automation, connectivity, customer interface)? With which market players should we join forces (strategic partnerships, -'co-opetition')? Which platforms/standardization processes should we play an active part in shaping? At what points should we exert political and market influences? How should we develop and improve our cyber-security? | Roadmap for digital transformation developed. |

*Source:* Own elaboration based on: *The Digital Transformation of Industry.* BDI, 2015.

industries are among the industries that are slower to implement Industry 4.0 technologies. The investment period is longer than in other industries, e.g., food industry. Moreover, the increase in investment is more visible in large capital groups, in large companies than in Small and Medium-Sized Enterprises (SMEs).

Wholesalers' markets are investing in new warehouse areas with full automation of activities: storage, movement, dispensing. The innovative equipment of the warehouses is: high-speed cranes for loading products, autonomous vehicles, cloud computing technologies, smart device sensors, electronic navigation systems, digital product codes in Product Lifecycle Management (PLM) (Myszor, 2018; Arora et al., 2008; Młody, 2018; Dziadkiewicz, 2015; Murri et al., 2019).

Individual markets are geared towards specialization or differentiation of production. Markets with personalized products are developing more dynamically than traditional markets. Customizing products to the individual needs of customers and taking care of their diversity are priorities of companies producing consumer goods. First, the customer is encouraged to reveal his preferences, and then the product is designed to suit him well. There are companies on the market that implement strategies of several variants of a given product so that the customer can choose the most attractive product for him. Diversification of products is observed in many sectors on the markets of individual customers, such as books, consumer electronics, toys, footwear, sports accessories, furniture, clothing (Dziadkiewicz, 2015).

Different industries are creating their own smart factory designs. In smart factories, production lines are prepared to make short or very short runs and make changes to products according to customer expectations. Fully automated lines run on modular systems, allowing them to quickly change the product mix. There are companies in the global market that are highly digitized, such as Boeing, General Electric, Adidas, Bosch, Uber, BMW, McDonald's, H&M, Henkel. Companies that have introduced deep technological innovations are benchmarks for other market participants, the same industries, or related industries. Innovation in companies takes place according to the specifics of a given market, type of production, and type of manufacturing technology. According to a report by McKinsey (2015), achieving high digital maturity in companies is determined by the structures of the networks to which companies belong. The following business network structures are mentioned in the McKinsey report:

- systems made up of very large companies (strong capital groups) with a high level of specialized technology and advanced manufacturing and process control, with a thousand sensors providing real-time online data to the decision-making systems of companies equipped with high-end decision-making tools (highly sophisticated);
- systems formed by large companies operating in industrial parks or business structures, with mass and personalized production (e.g., Adidas, Bosch);
- systems made up of small and medium-sized enterprises, with flexible technologies, delivering a variety of products to the immediate neighborhood (local customer, niche customer).

Based on PwC report entitled: *Industry 4.0: Building the Digital Enterprise* (2016) and the Springer book entitled: *Managing The Digital Transformation* (2018), and the manuscript (Geissbauer et al., 2016) six practical steps (Table 10) provide companies comprehensive perspective on Industry 4.0 by representing its own maturity model *Blueprint for Digital Success*:

- **Step 1:** Map out your Industry 4.0 strategy
- **Step 2:** Create initial pilot projects
- **Step 3:** Define the capabilities you need
- **Step 4:** Become a virtuoso in data analytics
- **Step 5:** Transform into a digital enterprise
- **Step 6:** Actively plan an ecosystem approach.

**Table 10**  The path of digital transformation.

| *Steps* | *Description* |
|---|---|
| **Step 1:** Map out your Industry 4.0 strategy | Analyze the current situation and create a strategy:<br>• check the condition of the machine park;<br>• analysis of IT system capabilities. |
| **Step 2:** Create initial pilot projects | • selection of IT technologies and service provider;<br>• establish the possibility of combining the old technology with the new one;<br>• establish the possibility of adding new modules (components) to already working process support information systems. |
| **Step 3:** Define the capabilities you need | • identification of financing sources for projects:<br>  ▪ profit or depreciation allowances;<br>  ▪ share or outside capital;<br>  ▪ share issue;<br>  ▪ bank credits;<br>  ▪ others. |
| **Step 4:** Become a virtuoso in data analytics | • implementation of procedures and data processing;<br>• advanced decision-making and process;<br>• optimization systems;<br>• autonomous robots and machine learning. |
| **Step 5:** Transform into digital enterprise Description | • integrated computer systems<br>• vertical and horizontal integration |
| **Step 6:** Actively plan an ecosystem approach | • create a digital enterprise culture;<br>• business process integration;<br>• digital leadership;<br>• digital cooperation with the outside world. |

*Source:* Own elaboration.

Planning for technological development begins with a diagnosis and determination of the possibility of introducing changes in the enterprise (the first step in Figure 2).

The starting point for introducing radical changes is the provisions in the enterprise development strategy—formal declaration of the members of the top management—about the adoption of a development path based on the idea of Industry 4.0. Strategies most often include the provision: "our company wants to be in Industry 4.0". In planning documents, of Industry 4.0 is a formal acceptance of new challenges by top management (the company's board of directors). It can be assumed that it is an expression of the willingness of executives (business owners) to invest in smart and high technologies that will make the enterprise smart (Gajdzik and Štverková, 2023). Thus, the strategic level of the enterprise is the first level that initiates change in the company (Schumacher et al., 2016). The scope of this level is presented in Table 11.

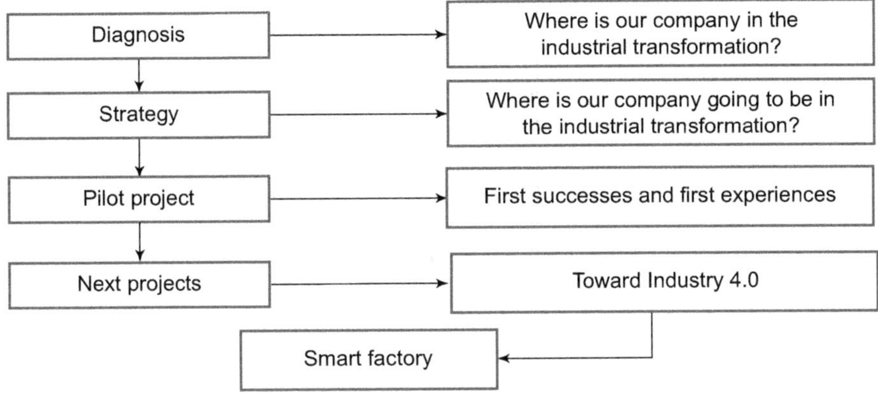


**Fig. 2**   Path of industrial transformation.
*Source*: Own elaboration, Gajdzik, 2022a Own study.

**Table 11**   Strategic level on the path of industrial transformation to digital maturity.

| Level | Key Question | Maturity |
| --- | --- | --- |
| Strategy | How would you describe the implementation status of your Industry 4.0 strategy? | • Pilot initiatives launched<br>• Strategy in development<br>• Strategy formulated<br>• Strategy in implementation<br>• Strategy implemented |

*Source:* Own elaboration based on Lichtblau et al., 2015.

At the planning stage, it is crucial to understand how new technologies fit into the business strategies. This is a task for production managers and management, but support from technology suppliers and integrators is also important. Their role in the preparation of Industry 4.0 solutions is multidimensional. Implementing new technologies often requires rethinking and adapting existing business models, which are challenges both at the strategic and operational levels. Enterprises in the transition to Industry 4.0, must not forget about the dynamic changes in the market, which require constant adaptation of business objectives and the complexity of integrating different systems and technologies. The analysis concludes by comparing technological aspects with economic benefits and social demands (Gajdzik, 2022c). In each of these areas, the factors likely to have a negative impact on the implementation of investment projects and the positive factors should be identified. At this stage, the company should also determine the risk factors for the planned investments and prepare control indicators. Risk management requires an integrated approach that takes into account both technological and organizational aspects. Early identification and management of risks can significantly increase the chances of project success.

Entering the direction of development based on new manufacturing and communication technologies into the strategy of the enterprise must be preceded by a diagnosis of the condition. To develop a strategy, it is necessary to determine the degree of technological advancement of the company and to set priorities for the future. It is necessary to work out a way of transforming the company, thanks to which it will be possible to use its strengths as the foundations of the planned changes. It is also necessary to determine their profitability.

At the beginning of their journey to full digital maturity in Industry 4.0, every company should answer the key diagnostic question: Where are they and where are they headed? (Gracel and Łebkowski, 2018). It is also worth considering the type of technology that can significantly help to achieve strategic goals and bring the company closer to the requirements of the new concept of industrial development. The company should perform a digital audit (Table 12). The following key questions are formulated during the audit (Głodek, 2006; Gajdzik, 2022a; Burgelman et al., 2004):

- What technologies and IT systems does the company use?
- What is the degree of digitization of processes?
- Do IT systems have interfaces to the leading system?
- What technologies are considered to be key to the company development?
- What are the company's opportunities to invest in the latest generation of IC technologies?

**Table 12**   Digital audit in industries.

| Technology | Key Question | Forms of Systems | | |
|---|---|---|---|---|
| Computer Systems | Which of the following systems do you use? | MES—manufacturing execution system<br>ERP—enterprise resource planning<br>PDM—product data management<br>PPS—production planning system<br>PDA—production data acquisition<br>MDC—machine data collection<br>CAD—computer-aided design<br>SCM—supply chain management | | |
| Interface to Leading System | | | Yes | No |
| Computer Systems | Does the system have an interface to the leading system? | MES—manufacturing execution system<br>ERP—enterprise resource planning<br>PDM—product data management<br>PPS—production planning system<br>PDA—production data acquisition<br>MDC—machine data collection<br>CAD—computer-aided design<br>SCM—supply chain management | | |

*Source:* Own elaboration based on Lichtblau et al. (2015).

In the process of changes, consultants from IC companies are invited to perform a digital audit, working closely with the management and staff of the audited enterprise. The result of the audit is the determination of the directions for the development of ICTs in the enterprise and the path of action. The directions of change determined, based on the digital audit, are presented to the management of the enterprise. The degree of digitization of the company is a determinant of further investment in high technologies. A low level of digitization of the company is a barrier, while a high level favors the transformation to smart manufacturing. Table 13 presents key opportunities for new technologies in the era of Industry 4.0. Companies implementing Industry 4.0 technologies, in the first phase, act as consultants, helping organizations understand what technologies will be most beneficial to them. They also create strategic implementation plans that align with the client's business goals. Many times, the integrator companies' activities are supported by technology providers, who help build so-called digital transformation roadmaps.

**Table 13**  Technological audit of Industry 4.0 pillars.

| *Capabilities* | *Technologies* |
|---|---|
| Real-time data management (Collection/Processing/Analysis/Inference) Interoperability Virtualization Decentralized Agility Integrated business processes | Data Analytics and Artificial Intelligence Adaptive Robotics Simulation Communication and Networks Cybersecurity Additive Manufacturing and 3D Printing Virtualization Sensors and Actuators Cloud Computing and Services RFID and RTLS Mobile Technologies |

*Source:* Based on: (Akdil et al., 2018; Ustundag and Cevikcan, 2018).

After the planning phase, time passes for implementation. Here, integrators adapt technologies to the specific needs of the customer and integrate them with existing systems (Gajdzik et al., 2024). This is not only a technical issue, but also an organizational one. Tests and validations are necessary to make sure everything works as expected. Technological projects are a collection of many simultaneous and mutually interacting investment and organizational operations. According to the management methodology, projects should be divided into technological and organizational (Baccarini, 1996; Gajdzik and Wolniak, 2022). The implementation of changes in companies starts with initial (pilot) investment programs, followed by subsequent ones. Pilot projects allowing to check whether the implemented solutions actually function as assumed at the stage of their design. While not every pilot project is a complete success, the knowledge gained during its implementation may prove invaluable.

To implement projects smoothly, it is necessary to develop a schedule for their implementation (Gajdzik et al., 2021a). Each project is tailored to the specifics of the execution area. Often, during the administration of projects, a feedback loop is made with verification of the assumptions made and the scope of changes, which most often need to be expanded and incorporated in subsequent projects. (Andersen and Jessen, 2003). The initiator of change in enterprises is the top management. When looking for a provider of solutions that we intend to implement in the enterprise, it is worth taking into account not only the price offered, but also comprehensiveness and compatibility, i.e., the possibility of any combination and supplementation of individual components of the system. It is also necessary to pay attention to the support of specialists, which is absolutely essential at the beginning of the road towards Industry 4.0.

Industrial investments are expensive, and not all companies have the spare funds to implement the high technologies. However, if it intends to undertake this through equity, it can use profit or depreciation here. It is also possible to increase share capital and use shareholder or stakeholder contributions. Issuing new shares may also be an appropriate solution. Usually, however, it is worthwhile to try to obtain financing for the investment from external funds. At the disposal of Polish companies are, among others, national and regional EU operational programs within the framework of grants from the National Center for Research and Development and the Operational Program Intelligent Development, programs for research and implementation, such as Horizon 2020, which is the largest ever EU program for scientific research and innovation. The development of a company can also be financed through foreign capital, i.e. loans, credits or debt securities such as bonds. The help of venture capital funds and angel business can also be a good idea.

Industrial investments are also stretched over time and investors, banks, and shareholders want to get a quick return on invested capital. Companies are diversifying their portfolio of projects. The set of projects implemented in enterprises includes projects that have been implemented for several months, as well as projects that have been implemented for several years (Table 14).

Technologies allow to obtain data that significantly improves production. However, to take full advantage of them, it is necessary to create data management procedures in the enterprise. It is also very important to choose the right platform to analyze them. Ideally, it should be one integrated solution for the entire organization. Companies develop and implement strategies to transform factories into smart factories. Acceptance of change at all levels of business and in all areas of activity, over time, has led these companies to create strong CPSs.

Organizational units responsible for supervizing the implementation of projects are appointed in companies implementing change projects. In companies, coordinators of individual projects are appointed, who are most often employees selected by the management board of the company. The management structure also

**Table 14** Schedules of investment projects.

| Projects | Time |
|---|---|
| Defining the company's development strategy with directions to I4.0 | a few months |
| Improvements at the IT level, including modifications used in the form of computer systems supporting processes. | 2 – 4 years |
| Creation of IT-computer links between individual branches of the group through compatible database management systems. | 2 years and longer |
| Installation of sensors on the production line | a year |
| Automation and robotization of the production line | 4 years and longer |
| Cybersecurity system | continuous |
| IT-computer links of production and logistics | 3 years and longer |
| Additive manufacturing: purchase and installation | a few months |
| A platform to process and visualize production data in real time | a few months |
| Installation of robots on the production line with full data analytics and real-time data delivery | 5 years and longer |
| Database with access to IoT and cloud computing services | a year and longer |
| Installation of RFID gates | a few months |
| System for automatic registration of transported parts of products | a year and longer |
| Information exchange systems between production and warehouse | a year and longer |
| Warehouse automation and robotization | 3 years and longer |
| Multidimensional analysis of data from all business processes | continuous |
| Cloud for customers (contractors) of the company with access to Big Data and IoT | a year |
| Dedicated IT systems for production management (any decision on the optimal use of resources supported by real-time data | 3 years |

*Source:* Own elaboration based on: (Gajdzik et al., 2021a; Gajdzik, 2022a.)

includes a board representative for digitalization of the company. Companies use the term Chief Digital Officer (CDO), Chief Digital Information Officer (CDIO) for the created positions. Branch units are responsible for the implementation of individual projects. At senior or middle management levels, key projects are overseen by managers with different control areas. The management team, together with project coordinators, participates in directing and supervizing individual activities and managing processes related to the implemented investment projects. Implementing Industry 4.0 technology is a collaborative process. First, cooperation between departments within the organization is essential. Departments such as IT, Manufacturing, Maintenance, Quality Control, Management, and Finance must work together to be successful and collaborate with the system integrator.

Cooperation should be the cornerstone of the implementation of projects in the company. We also need clear leadership, committed managers setting goals and supporting subordinates (Kwiotkowska et al., 2021). At the same time, it is necessary to teach engineers and operational staff business thinking, that is, to translate projects that are to be implemented into business indicators that show what added value a given initiative generates for the company. It is worth taking advantage of the development opportunities they offer and start implementing them in your company as soon as possible.

The challenge for manufacturing companies is to integrate Industry 4.0 technology into the existing structure (old technology). It is a process that can be divided into three main categories: technical, organizational, and strategic. When it comes to technical aspects, the first step is to understand the compatibility of the systems. Often, existing IT systems and production machines are incompatible with new technologies. The second element is security. New technologies can introduce additional attack vectors, so security is critical. The third element is the scalability of the systems necessary for future development. When it comes to organizational aspects, we are talking about changing organizational culture, change management and training, and competence development. The implementation of new technologies affects the culture of the company and requires change management at different levels of the organization. In the context of strategic aspects, it is crucial to understand how new technologies fit into an organization's business goals and how is the Return On Investment (ROI). The ultimate success of new solutions depends on whether the costs are balanced with the potential benefits. In addition, project management must maintain regulatory compliance, which is key to successful integration.

Once new technologies and systems are implemented, third-party companies train employees to use the new technologies effectively. They also provide technical support and help with data analysis, which can lead to further process optimization. Integrating Industry 4.0 technologies is a complex process that requires well-coordinated efforts at many levels of the organization. Proper planning, management and management support are absolutely critical to the success of this endeavor.

To sum up, digital transformation of companies is driven by investments in telecommunication technologies and high technologies. In this process, the need of the integration in current technologies with new machines and automated work processes to obtain horizontal and vertical integration along the value chain (Gajdzik, 2022b). The implementation of the Industry 4.0 strategy is based on a number of projects that build (step by step) smart manufacturing.

## References


Abdollahzadegan, A., Che Hussin, A.R., Moshfegh Gohary, M. and Amini, M. 2013. The organizational critical success factors for adopting cloud computing in SMEs. *J. Inf. Syst. Res. Innov. (JISRI)*, *4*: 67–74.

Aheleroff, S., Xu, X., Lu, Y., Aristizabal, M., Velásquez, J.P., Joa, B. and Valencia, Y. 2020. IoT-Enabled Smart Appliances under Industry 4.0: A Case Study. *Advanced Engineering Informatics*, 101043.

Akyurt, İ.Z., Kuvvetliand, Y. and Deveci, M. 2021. Enterprise Resource Planning in the Age of Industry 4.0: A General Overview, Chapter 13, *In*: Turan Paksoy, Çiğdem Koçhan, Sadia Samar Ali (Eds.), *Logistics 4.0 Digital Transformation of Supply Chain Management*. Raylor and Franics Group.

Akdil K.Y., Ustundag, A. and Cevikcan E. 2018 (January). Maturity and Readiness Model for Industry 4.0 Strategy, Chapter 4, *In*: Ustundag, A. Cevikcan, E. *Industry 4.0: Managing the Digital Transformation*. Springer. DOI: 10.1007/978-3-319-57870-5.

AlBar, A.M. and Hoque, M.R. 2017. Factors affecting cloud ERP adoption in Saudi Arabia: An empirical study. *Inf. Dev.*, *35*: 150–164.

Andersen, E.S. and Jessen, S.A. 2003. Project maturity in organizsations. *International Journal of Project Management*, *21*(6): 457–461.

Arora, N., Dreze, X., Ghose, A., Hess, J.D., Iyengar, R., Bing, J., Joshi, Y., Kumar, V., Lurie, N., Neslin, S., Sajeesh, S., Su, M., Niladri, S., Thomas, J. and Zhang, Z.J. 2008. Putting one-to-one marketing to work: Personalization, customization, and choice. *Springer Science and Business Media*, 310.

Aslam, J., Saleem, A., Khan, N.T. and Kim, Y.B. 2021. Factors influencing blockchain adoption in supply chain management practices: A study based on the oil industry. *J. Innov. Knowl.*, *6*: 124–134.

ASTOR, 2016. *Industry 4.0 Whitepaper*. Available online: www.astor.com.pl/industry4: Industry 4.0.Authors: T. Iwański, J. Gracel Whitepaper, Kraków. [Online], www.astor.com.pl/industry4. (Access data: 2020-05-19).

Aurelie, D. 2008 (June). The alignment between customer relationship management and IT strategy: A proposed research model. *In*: *Proceedings of the Southern Association for Information Systems Conference*, Richmond, VA, USA, 13 March–15 June 13–18.

Baccarini, D. 1996. The concept of project complexity: A review. *International Journal of Project Management*, *14*(4): 201–204.

Bai, C., Dallasega, P., Orzes, G. and Sarkis, J., 2020. Industry 4.0 technologies assessment: A sustainability perspective. *Int. J. Prod. Econ.*, *229*: 107776. https://doi.org/10.1016/ j.ijpe.2020.107776.

Banerjee, A. 2019. Blockchain with IOT: Applications and use cases for a new paradigm of supply chain driving efficiency and cost. *In*: *Advances in Computers* (1st Edn.), Vol. 115. Elsevier: Amsterdam, The Netherlands.

Benias, N. and Markopoulos, A.P. 2017 (September). A review on the readiness level and cyber-security challenges in Industry 4.0. *In*: *Proceedings of the 2017 South Eastern European Design Automation, Computer Engineering, Computer Networks and Social Media Conference* (*SEEDA-CECNSM*), Kastoria, Greece, 23–25. IEEE: New York, NY, USA.

BCG: Boston Consulting Group (Hg.) 2015. *Industry 4.0: The Future of Productivity and Growth in Manufacturing Industries*. München, Germany.

BDI, 2015. *The Digital Transformation of Industry*. A European study commissioned by the Federation of German Industries (BDI) and conducted by Roland Berger Strategy Consultants. Berlin. https:// www.rolandberger.com › publications.

Bonomi, F., Milito, R., Zhu, J. and Addepalli, S. 2012 (August) Fog Computing and Its Role in the Internet of Things. *In*: *Proceedings of the First Edition of the MCC Workshop on Mobile Cloud Computing*, 13–16. Helsinki, Finland, 17 August.

Bortolini, M., Ferrari, E., Gamberi, M., Pilati, F. and Faccio, M., 2017. Assembly system design in the industry 4.0 era: A general framework. *IFAC-PapersOnLine*, *50*: 5700–5705.

Bouwman, H., de Reuver, M. and Nikou, S. 2017. The impact of digitalization on business models: How IT artefacts, social media, and Big Data force firms to innovate their business model. *In*: *Proceedings of the 14th Asia-Pacific Regional Conference of the International Telecommunications Society (ITS): Mapping ICT into Transformation for the Next Information Society*, Kyoto, Japan, 24–27 June. 20. Boyes, H., Hallaq, B., Cunningham, J. and Watson, T. 2018. The industrial internet of things (IIoT): An analysis framework. *Computers in Industry*, *101*: 1–12.

Burgelman, R.A., Christensen, C.M. and Wheelwright, S.C. 2004. *Strategic Management of Technology and Innovation*. McGraw-Hill, New York.

Burrell, D. 2019. *Principles of Industry 4.0 and the 9 Pillars*. [Online:] https://www.plextek.com/ insights/insights-insights/industry-4-0-and-the-9-pillars/ (Access data: 2019-02-07).

Castagnoli, R., Büchi, G., Coeurderoy, R. and Cugno, M. 2021. Evolution of industry 4.0 and international business: A systematic literature review and a research agenda. *Eur. Manag. J.,* *40*: 572–589.

Carvalho, H., Barroso, A.P., Machado, V.H., Azevedo, S. and Cruz-Machado, V. 2012. Supply chain redesign for resilience using simulation. *Comput. Ind. Eng., 62*: 329–341.

Chearavanont, S. 2020. *How Digitization and Innovation Can Make the Post-Covid World a Better Place.* World Economic Forum. https://www.weforum.org/agenda/2020/08/how-digitization-and-innovation-can-make-thepost-covid-world-a-betterplace/.

Chowdhury, S., Budhwar, P., Dey, P.K., Joel-Edgar, S. and Abadie, A. 2022. AI-employee collaboration and business performance: Integrating knowledge-based view, socio-technical systems, and organizational socialization framework. *J. Bus. Res., 144*: 31–49.

Ciffolilli, A. and Muscio, A. 2018. Industry 4.0: National and regional comparative advantages in key enabling technologies. *European Planning Studies, 26*(12): 2323–2343.

Csalódi, R., Süle, Z., Jaskó, Sz., Holczinger, T. and Abonyi, J. 2021. Industry 4.0-driven Development of Optimization Algorithms: A Systematic Overview. *Hindawi,* Review Article, Open Access, Article ID 6621235. https://doi.org/10.1155/2021/6621235.

Cui, L., Gao, M., Dai, J. and Mou, J. 2020. Improving supply chain collaboration through operational excellence approaches: An IoT perspective. *Ind. Manag. Data Syst., 122*: 565–591.

Culot, G., Nassimbeni, G., Orzes, G. and Sartor, M. 2020. Behind the definition of industry 4.0: Analysis and open questions. *International Journal of Production Economics, 226*: 107617.

Dai, B. and Liang, W. 2022. The Impact of Big Data Technical Skills on Novel Business Model Innovation Based on the Role of Resource Integration and Environmental Uncertainty. *Sustainability, 14*: 2670.

Danel, R., Gajdzik, B. and Ropyak, L. 2023. *Trends in Data Processing Control in Continuous Production Systems.* Zeszyty Naukowe Politechniki Śląskiej. Organization and Management, Silesian University of Technology. *181*: 93–105. DOI:10.29119/1641-3466.2023.181.6.

Dash, R., McMurtrey, M., Rebman, C. and Kar, U.K. 2019. Application of artificial intelligence in automation of supply chain management. *J. Strateg. Innov. Sustain., 14*: 43–53.

Definition of automation from the *Cambridge Academic Content Dictionary.* Cambridge University Press.

Deloitte, 2020. *Industry 4.0 in Automotive.* https://www2.deloitte.com/content/dam/insights/us/articles/automotive-news_industry-4-0-in-automotive/DI_Automotive-News-Supplement.pdf.

Deloitte Insights. 2018. *The Industry 4.0 Paradox. Overcoming Disconnects on the Path to Digital Transformation. Deloitte Development LLC.* [Online:] https://www2.deloitte.com/content/dam/Deloitte/cn/Documents/energy-resources/deloitte-cn-er-industry-4.0-paradox-overcoming-disconnects-en-full-report-190225.pdf (Access data: 2020-09-09).

Dev, N.K., Shankar, R. and Swami, S. 2020. Diffusion of green products in industry 4.0: Reverse logistics issues during design of inventory and production planning system. *International Journal of Production Economics, 223*: 107519.

Dopico, M., Gomez, A., De la Fuente, D., García, N., Rosillo, R., J. Puche, J. 2016. A vision of industry 4.0 from an artificial intelligence point of view. *In: Proceedings on the International Conference on Artificial Intelligence (ICAI),* 407. The Steering Committee of the World Congress in Computer Science, Computer Engineering and Applied Computing (WorldComp).

Drath, R. and Horch, A. 2014. Industrie 4.0: Hit or hype? *IEEE Industrial Electronics Magazine, 8*(2): 56–58.

Dubey, G., Gupta, R. K., Kumar, S. and Kumar M. 2022. Study of Industry 4.0 pillars and their uses in increasing productivity and reducing logistics defects. *Materials Today Proceedings, 63*: 85–91. https://doi.org/10.1016/j.matpr.2022.02.335.

Dziadkiewicz, A. 2017. Personalizacja a kastomizacja w marketing. *Zarządzanie i Finanse. Journal of Management and Finance, 15*(1): 95–108.

Eliasy, A. and Przychodzen, J. 2020. The role of AI in capital structure to enhance corporate funding strategies. *Array, 6*: 100017.

Elkington J. 2020 (April 7). *Green Swans: The Coming Boom in Regenerative Capitalism* (Hardcover) Fast Company Pr., USA.

Erboz, G. 2017 (November). How to Define Industry 4.0: Main Pillars of Industry 4.0. *In: 7th International Conference on Management* (*ICoM 2017*), Nitra, Slovakia, [Online:] https://www.researchgate.net/publication/326557388_How_To_Define_Industry_40_Main_Pillars_Of_Industry_40 (Access data: 2019-06-09).

European Commission. 2015. *The Factories of the Future.* [Online] Available from: http://www.effra.eu/attachments/article/129/Factories%20of%20the%20Future%202020%20Roadmap.pdf (Access data: 2015-07-10).

European Parliament. 2016. *Industry 4.0: Digitalization for Productivity and Growth.* [Online] Available from: www.europarl.europa.eu/RegData/etudes/BRIE/.../EPRS_BRI(2015)568337EN.pdf. (Access data: 2016-09-10).

European Commission: *e-Skills for Growth and Jobs.* [Online] Available at: http://ec.europa.eu/growth/sectors/digital-economy/e-skills/.

European document: *Industry 5.0: Towards a Sustainable, Human-centric and Resilient European Industry. Report.* European Commission (Directorate-General for Research and Innovation (European Commission) 4 January, 2021, esearch-and-innovation.ec.europa.eu/knowledge-publications-tools-and-data/publications/all-publications/industry-50-towards-sustainable-human-centric-and-resilient-european-industry_en.

European document: *Industry 5.0. Towards a Sustainable, Human-centric, and Resilient European Industry.* (R&I PAPER SERIES POLICY BRIEF – Research and Innovation; European Commission Directorate – General for Research and Innovation Directorate F, Prosperity Unit F.5, Industry 5.0) and literature review.

European document: *Industry 5.0: A Transformative Vision for Europe. Governing Systemic Transformations towards a Sustainable Industry.* ESIR Policy Brief No. 3. (European Commission, Directorate – General for Research and Innovation, Directorate G – Common Policy Centre, Unit G1 – Common R&I Strategy & Foresight Service).

Fatorachian, H. and Kazemi, H. 2018. A critical investigation of Industry 4.0 in manufacturing: theoretical operationalization framework. *Production Planning and Control, 29*(8): 633–644.

Feliciano-Cestero, M.M., Ameen, N., Kotabe, M., Paul, J. and Signoret, M. 2023. Is digital transformation threatened? A systematic literature review of the factors influencing firms' digital transformation and internationalization. *J. Bus. Res., 157*: 113546.

Fisher, O., Watson, N., Porcu, L., Bacon, D., Rigley, M. and Gomes, R.L. 2018. Cloud manufacturing as a sustainable process manufacturing route. *J. Manuf. Syst., 47*: 53–68.

Frank, A.G., Dalenogare, L.S. and Ayala, N.F. 2019. Industry 4.0 technologies: implementation patterns in manufacturing companies. *International Journal of Production Economics, 210*: 15–26.

Gajdzik, B. 2020. Development of business models and their key components in the context of Cyber-Physical Production Systems in Industry 4.0. Chapter 3, *In: A. Jabłoński, M. Jabłoński, (Eds.), Scalability and Sustainability of Business Models in Circular, Sharing and Networked Economies,* 73–94. Cambridge Scholars Publishing. ISBN (10): 1-5275-4609-8, ISBN (13): 978-1-5275-4609-7.

Gajdzik, B. 2022a. *Diagnoza kierunków transformacji przemysłu stalowego w Przemyśle 4.0,* Title in English: *Diagnosis of the Steel Industry Transformation Direction in the Industry 4.0.* Monograph 945: Silesian University of Technology, Gliwice.

Gajdzik, B. 2022b. Frameworks of the Maturity Model for Industry 4.0 with Assessment of Maturity Levels on the Example of the Segment of Steel Enterprises in Poland. *Journal of Open Innovation: Technology, Market, and Complexity.* 2022; *8*(2): 77. https://doi.org/10.3390/joitmc8020077.

Gajdzik, B. 2022c. The technological cluster in business model of steel production with the list of determinants of the development to level 4.0: STEEPVL analysis. *Proceedings of the 39th International Business Information Management Association Conference* (*IBIMA*), May 2022.

Gajdzik, B. 2022d. Towards Industry 4.0 versus the COVID-19 crisis - the use of selected technologies (ICT) in business, *In:* Semwal T., Iqbal F. (Eds.), *Cyber-physical Systems Solutions to Pandemic Challenges,* 53–70. Taylor & Francis Group, ISBN 978-1-032-03037-1.

Gajdzik, B. 2023a. Industry 5.0 as a New Concept of Development Within High Volatility Environment: About the Industry 5.0 Based on Political and Scientific Studies. Silesian University of Technology Publishing House Scientific Papers of Silesian University of Technology. *Organization and Management Series, 169*: 255–279.

Gajdzik, B. 2023b. Industry 5.0 as the upgrade of industry 4.0: Towards one common concept of industrial transformation. Silesian University of Technology Publishing House Scientific Papers of Silesian University of Technology. *Organization and Management Series, 181*: 131–150.

Gajdzik B., Grabowska S. and Saniuk S. 2021a. A Theoretical Framework for Industry 4.0 and Its Implementation with Selected Practical Schedules. *Energies, 14*(4), 1–24, 940. https://doi.org/ 10.3390/en14040940, doi: 10.3390/en14040940.

Gajdzik, B., Grabowska, S. and Saniuk, S. 2021b. Key socio-economic megatrends and trends in the context of the Industry 4.0 framework. *Forum Scientiae Oeconomia, 9*(3), 5–22. https://doi. org/10.23762/FSO_VOL9_NO3_1.

Gajdzik, B., Grabowska, S. and Wyciślik, A. 2019. Explanatory preview of directions of changes in development of Industry 4.0. *Polish Technical Review, 1*: 5–9. DOI: 10.15199/180.2019.1.1.

Gajdzik, B., Siwiec D., Wolniak, R. and Pacana, A. 2024. Approaching Open Innovation in Customization Frameworks for Product Prototypes with Emphasis on Quality and Life Cycle Assessment (QLCA). *Journal of Open Innovation: Technology, Market, and Complexity, 10*(20): 1–16. DOI: 10.1016/j.joitmc.2024.100268.

Gajdzik, B. and Štverková, H. 2023. Frameworks of strategic management in Industry 4.0. *W: Proceedings of the 15th International Conference on Strategic Management and Its Support by Information Systems, 72–85*. May 22–24. Ostrava, Czech Republic/Němec Radek, Chytilová Lucie (red.), 2023, VŠB – Technical University of Ostrava.

Gajdzik, B. and Wolniak, R. 2022 (June). Smart Production Workers in Terms of Creativity and Innovation: The Implication for Open Innovation. *Journal of Open Innovation: Technology, Market, and Complexity, 8*(2):68., https://doi.org/10.3390/joitmc8020068.

Gartner: *Digitalization*. https://www.gartner.com/en/information-technology/glossary/digitalization.

Geissbauer, R., Vedso, J. and Schrauf, S. 2016. *Industry 4.0: Building the Digital Enterprise*. Retrieved from PwC Website: https://www.pwc.com/gx/en/industries/industries-4.0/landing-page/ industry-4.0-building-your-digital-enterprise-april-2016.pdf.

Ghobakhloo, M. and Iranmanesh, M. 2021. Digital transformation success under Industry 4.0: A strategic guideline for manufacturing SMEs. *Journal of Manufacturing Technology Management, Emerald*. DOI: 10.1108/JMTM-11-2020-0455.

Głodek, P. 2006. *Transfer technologii do małych i średnich przedsiębiorstw. Vademecum innowacyjnego przedsiębiorcy*. Tom 1, STIM, SOOIPP, Warszawa.

Grabowska, S., Saniuk, S. and Gajdzik, B. 2022. Industry 5.0: Improving humanization and sustainability of Industry 4.0. *Scientometrics, 127*: 3117–3144, https://doi.org/10.1007/s11192-022-04370-1.

Gracel, J. and Łebkowski, P. 2018. Concept of Industry 4.0-Related Manufacturing Technology Maturity Model (ManuTech Maturity Model – MTMM) Decision Making. *Manufacturing and Services, 12* (1–2): 17–31. [Online]: https://journals.agh.edu.pl/dmms/article/view/2779/2193.

Greengard, S. 2015. *The Internet of Things*. Boston, MA: MIT Press.

Groover M.P. 2024. The Editors of Encyclopaedia Britannica Last Updated: Jan. 19, 2024 Article History. https://www.britannica.com/technology/automation/additional-info#history.

Grossman, R.L., Gu, Y., Sabala, M. and Zhang, W.2009. Compute and storage clouds using wide area high performance networks. *Future Gener. Comput. Syst., 25*: 179–183.

Gupta, S., Modgil, S. and Gunasekaran, A. 2020. Big data in lean six sigma: A review and further research directions. *Int. J. Prod. Res., 58*: 947–969.

Hallikainen, H., Savimäki, E. and Laukkanen, T. 2020. Fostering B2B sales with customer big data analytics. *Ind. Mark. Manag., 86*: 90–98.

Hermann, M., Pentek, T. and Otto, B. 2016. Design Principles for Industrie 4.0 Scenarios. *49th Hawaii International Conference on System Sciences*.

Hofmann, E. and Rüsch, M., 2017. Industry 4.0 and the current status as well as future prospects on logistics. *Comput. Ind.*, *89*: 23–34. IEA Council, 2019.

Ilcus, A.M. 2018. Impact of Digitalization in Business World. *Rev. Manag. Comp. Int.*, *19*: 350–358.

IFR. *World Robotics Report 2022*.

ITU. 2023. *Measuring Digital Development: Facts and Figures 2023*. The International Telecommunication Union's (ITU). https://www.itu.int/en/ITU-D/Statistics/Pages/facts/default.aspx.

Ivanov, D. and Dolgui, A. 2019. New disruption risk management perspectives in supply chains: Digital twins, the ripple effect, and resilieanness. *IFAC-PapersOnLine*, *52*: 337–342.

Jafari, N., Azarian, M. and Yu, H. 2022. Moving from Industry 4.0 to Industry 5.0: What are the Implications for Smart Logistics? *Logistics*, *6*: 26.

Jain, V., Kumar, S., Soni U., and Chandra, C. 2017. Supply chain resilience: Model development and empirical analysis. *Int. J. Prod. Res.*, *55*: 6779–6800.

Kagermann H., 2014a. *Change Through Digitization—Value Creation in the Age of Industry 4.0. Management of Permanent Change*, 23–45. Springer Gabler, Wiesbaden. DOI: https://doi.org/10.1007/978-3-658-05014-6_2.

Kagermann, H. 2014b. Chancen von Industrie 4.0 nutzen (Seizing opportunities of Industry 4.0). *In*: T. Bauernhansl, M. Hompel, B. Vogel-Heuser (Eds.), *Industrie 4.0 in Produktion, Automatisierung, und Logistik. Anwendung, Technologien, und Migration* (*Industry 4.0 in Production, Automation, and Logistics. Application, Technologies and Migration*), 603–614. Springer. Wiesbaden, Germany.

Kagermann, H. 2015. Change through digitization: Value creation in the age of Industry 4.0. *In*: Albach, H., Meffert, H., Pinkwart, A., Reichwald, R. (Eds.), *Management of Permanent Change*, 23–45. Springer, Fachmedien, Wiesbaden. Available at: https://link.springer.com/ book/10.1007/978-3-658-05014-6#editorsandaffiliations.

Kagermann, H., Helbig, J., Hellinger, A. and Wahlster, W. 2013. *Recommendations for Implementing the Strategic Initiative INDUSTRIE 4.0: Securing the Future of German Manufacturing Industry*. Final Report of the Industrie 4.0 Working Group, Forschungsunion.

Kumar, S.L. 2017. State of the art-intense review on artificial intelligence systems application in process planning and manufacturing. *Engineering Applications of Artificial Intelligence*, *65*: 294–329.

Kumar, R., Singh, R.K. and Dwivedi, Y.K. 2020. Application of Industry 4.0 technologies in Indian SMEs for sustainable growth: Analysis of challenges. *Journal of Cleaner Production*, *257*: 124063.

Kumar, K., Zindani, D. and Davim, J.P. 2019. *Digital Manufacturing and Assembly Systems in Industry 4.0*. CRC Press: Boca Raton, FL, USA.

Kurfess, T.R., Saldana, Ch., Saleeby, K. and Dezfouli, M.P. 2020 (Nov.). A Review of Modern Communication Technologies for Digital Manufacturing Processes in Industry 4.0 *J. Manuf. Sci. Eng.*, *142*(11): 110815 (8 pages), Paper No: MANU-19-1687. https://doi.org/10.1115/1.4048206. Published Online: September 28, 2020.

Kwiotkowska, A., Gajdzik, B., Wolniak, R., Vveinhardt, J. and Gębczyńska, M. 2021. Leadership Competencies in Making Industry 4.0 Effective: The Case of Polish Heat and Power Industry. *Energies*, *14*: 4338. https://doi.org/10.3390/en14144338.

Lasi, H., Fettke, P., Kemper, H.G., Feld, T. and Hoffmann, M. 2014. Industry 4.0. *Business and Information Systems Engineering*, *6*(4): 239–242.

Law, Ch. C.H. 2019. *Managing Enterprise Resource Planning, Adoption, and Business Processes: A Holistic Approach*. Cambridge Scholars Publishing.

Leiting, A.K., De Cuyper, L. and Kauffmann, C. 2022. The Internet of Things and the case of Bosch: Changing business models while staying true to yourself. *Technovation*, *118*: 102497.

Lezzi, M., Lazoi, M. and Corallo, A. 2018. Cybersecurity for Industry 4.0 in the current literature: A reference framework. *Comput. Ind.*, *103*: 97–110.

Lichtblau, K., Stich, V., Bertenrath, R., Blum, M., Bleider, M., Millack, A.,... Schröter, M. 2015. *IMPULS-Industrie 4.0-Readiness. Impuls-Stiftung des VDMA*. Aachen-Köln.

Liu, Y., Peng, Y., Wang, B., Yao, S. and Liu, Z. 2017 (Jan. 16). Review on cyber-physical systems. *IEEE/CAA Journal of Automatica Sinica*, *4*(1) (January 2017): 27 – 40. DOI: 10.1109/JAS.2017.7510349.

Longo, F., Padovano, A. and Umbrello, S. 2020. Value-Oriented and Ethical Technology Engineering in Industry 5.0: A Human-Centric Perspective for the Design of the Factory of the Future. *Applied Sciences-BASEL*, *10*(12). DOI: 10.3390/app10124182.

Lorentz, M., Rüsmann, M. Strack, R., Lueth, K.L. and Bolle, M. 2015. *Man and Machine in Industry 4.0: How Will Technology Transform the Industrial Workforce through 2025?* [Online] Available at: https://www.bcgperspectives.com/content/articles/technology-businesstransformationengineered-products-infrastructure-man-machine-industry-4. 2015. [Access data: 2017-03-01].

Lou, P., Liu, Q., Zhou, Z. and Wang, H. 2011. Agile supply chain management over the Internet of Things. *In: Proceedings of the 2011 International Conference on Management and Service Science*, 1–4. Wuhan, China, 12–14 August.

Lu, Y. 2017. Industry 4.0: A Survey on Technologies, Applications, and Open Research Issues, *Journal of Industrial Information Integration*, *6*: 1–10.

Maddikunta, P.K.R., Pham, Q.-V., Prabadevi, B., Deepa, N., Dev, K., Gadekallu, T.R., Ruby, R. and Liyange, M. 2021. Industry 5.0: A survey on enabling technologies and potential applications. *J. Ind. Inf. Integr.*, *26*: 100257.

Manda, M.I. and Backhouse, J. 2017 (July). Digital transformation for inclusive growth in South Africa: Challenges and opportunities in the 4th industrial revolution. *African Conference on Information Systems & Technology (ACIST)* – Cape Town, South Africa, 10th–11th July.

Matthews, K. 2018. *What is Manufacturing 5.0?* Jun. 28. Retrieved from: https://www.manufacturing.net/software/article/13123438/what-is-manufacturing-50.

McKinsey, 2015. *Industry 4.0: How to Navigate Digitization of the Manufacturing Sector*. Report, McKinsey Company. [online:] https://www.mckinsey.com/~/media/McKinsey/Business%20Functions/Operations/Our%20Insights/Industry%2040%20How%20to%20navigate%20digitization%20of%20the%20manufacturing%20sector/Industry-40-How-to-navigate-digitization-of-the-manufacturing-sector.pdf. (Access data: 2020-06-10).

McMillan, G.K. and Vegas, P.H. 2019. *Process/Industrial Instruments and Controls*. Handbook (6th Edn.) McGraw-Hill Education.

Młody, M. 2018. Personalizacja produktów a Przemysł 4.0 – ocena słuszności implementacji nowoczesnych technologii w przemyśle produkcyjnym z perspektywy konsumentów. *Ekonomika i Organizacja Przedsiębiorstw*, *2*:, 62–72.

Murri, M., Streppa, E., Colla, V., Fornai, T. and Branca, A. 2019. *Digital Transformation in European Steel Industry: State of Art and Future Scenario 2022*, 1–43. Springer.

Müller, J. 2019. Business model innovation in small- and medium-sized enterprises: Strategies for Industry 4.0 providers and users. *J. Manuf. Technol. Manag.*, *30*: 1127–1142.

Myszor, P. 2018. Przemysł 4.0 w hutnictwie – personalizacja produktów stalowych przyszłością branży. *Wydawnictwo Nowy Przemysł*. 2018-10-30. [Online:] https://www.wnp.pl/artykuly/przemysl-4-0-w-hutnictwie-personalizacja-produktow-stalowych-przyszloscia-branzy,333103.html.

Naglic, A., Tominc, P. and Logožar, K. 2020. The impact of industry 4.0 on export market orientation, market diversification, and export performance. *Organizacija*, *53*: 227–244.

Nagy, J., Oláh, J., Erdei, E., Máté, D. and Popp, J. 2018. The role and impact of Industry 4.0 and the internet of things on the business strategy of the value chain: The case of Hungary. *Sustainability*, *10*: 3491.

Nahavandi, S. 2019. Industry 5.0: A Human-Centric Solution. *Sustainability*, *11*: 4371. DOI:10.3390/su11164371.

Neubert, M. 2019. The impact of digitalization on the speed of internationalization of lean global start-ups. *Technol. Innov. Manag. Review*, *8*: 44–54.

Ng, I., Scharf, K., Pogrebna, G. and Maull, R. 2015. Contextual variety, Internet-of-Things and the choice of tailoring over platform: Mass customization strategy in supply chain management. *Int. J. Prod. Econ.*, *159*: 76–87.

OECD, 2016. *Inclusive Growth*. Retrieved from http://www.oecd.org/inclusive-growth/.

Osterrieder, P., Budde, L. and Friedli, T. 2020. The smart factory as a key construct of industry 4.0: A systematic literature review. *Int. J. Prod. Econ., 221*: 107476.

Peters, H. 2016. Application of Industry 4.0 concepts at steel production from an applied research perspective. *Presentation at 17th IFAC Symposium on Control, Optimization, and Automation in Mining, Mineral, and Metal Processing.* [Online:] https://tc.ifac-control.org/6/2/files/symposia/vienna-2016/mmm2016_keynotes_peters.PowerPoint-Präsentation (ifac-control.org).

Peters, H. 2017. How could Industry 4.0 transform the Steel Industry? *Presentation at Future Steel Forum*, Warsaw, 14–15.6.2017. [Online:] PowerPoint-Präsentation (futuresteelforum.com). https://futuresteelforum.com/content-images/speakers/Prof.-Dr-Harald-Peters-Industry-4.0-transform-the-steel-industry.pdf.

Pinciroli, L., Baraldi, P. and Zio, E. 2023. Maintenance optimization in Industry 4.0. *Reliability Engineering and System Safety, 234*: 109204. https://doi.org/10.1016/j.ress.2023.109204.

Pizoń, J. and Gola, A. 2023. Human–Machine Relationship: Perspective and Future Roadmap for Industry 5.0 Solutions. *Machines, 11*: 203. https://doi.org/10.3390/machines11020203.

Pollak, A. 2022. *APA Group, NAZCA 4.0. Przemysł 4.0 w praktyce.* https://open.spotify.com/show/4lJapGvA6aBm7QNbWIKoWe Dostęp: 24.11.2022SS). *IEEE*, 3928–3937.

Polman, P. and Winston, A. 2021. Net Positive: How Courageous Companies Thrive by Giving More Than They Take. *Harvard Business Review* (Sept.–Oct. Issue).

Przemysł Przyszłości: https://elearning.przemyslprzyszlosci.gov.pl/przemysl-5-0-co-zmienia-w-zakresie-podejscia-do-procesu-transformacji/ (Access data: 02.02.2024).

PwC. 2022. Digital factory transformation survey. https://theonliner.ch/uploads/heroes/pwc-digital-factory-transformation-survey-2022.pdf.

PwC. Digital Factory Transformation Survey 2022. *PwC Study: Investments are Booming, but Implementation is Lagging Behind.* https://www.pwc.de/en/strategy-organisation-processes-systems/operations/digital-factory-transformation-survey-2022.html (Access data: 2022-11-24).

PwC, 2016. *Global Industry 4.0 Survey. What We mean by Industry 4.0/Survey key findings / Blueprint for Digital Success.* [Online:] https://www.pwc.com/gx/en/industries/industries-4.0/landing-page/industry-4.0-building-your-digital-enterprise-april-2016.pdf. (Access data: 2020-08-20).

Quetti, C., Pigni, F. and Clerici, A. 2012. Factors affecting RFID adoption in a vertical supply chain: The case of the silk industry in Italy. *Prod. Plan. Control, 23:* 315–331.

Robotisation:https://www.collinsdictionary.com/dictionary/english/robotisation#google_vignette.

Roblek V., Meško, M. and Krapež, A. April-June 2016. A Complex View of Industry 4.0. *SAGE Open*, 1–11. DOI: 10.1177/2158244016653987sgo.sagepub.com.

Romero, D., Stahre, J., Wuest, T., Noran, O., Bernus, P., Fast-Berglund, Å. and Gorecky, D., 2016. Towards an operator 4.0 typology: A human-centric perspective on the fourth industrial revolution technologies. *In: Proceedings of the International Conference on Computers and Industrial Engineering (CIE46)*, Tianjin, China.

Ross, Ph. and Maynard K. 2021. Towards a 4th industrial revolution. *Intelligent Buildings International, 13*(3): 159–161. DOI: 10.1080/17508975.2021.1873625.

Rüßmann, M., Lorenz, M., Gerbert, P., Waldner, M., Justus, J., Engel, P. and Harnisch, M. 2015. *Industry 4.0.* The Boston Consulting Group (BCG). http://image-src.bcg. com/Images/BCG-Przemysl-4-PL_tcm78-123996.pdf.

Rüßmann, M., Lorenz, M., Gerbert, P., Waldner, M., Justus, J., Engel, P. and Harnisch, M. 2015. *Industry 4.0. The Future of Productivity and Growth in Manufacturing Industries.* The Boston Consulting Group (Issue April).

Saniuk, S., Grabowska, S. and Gajdzik, B. 2020. Social expectations and market changes in the context of developing the industry 4.0 concept. *Sustainability, 12*: 1362.

Santos-Pereira, C., Durão, N., Moreira, F. and Veloso, B. 2022. The importance of digital transformation in international business. *Sustainability, 14*: 834.

Saldivar, A.A.F. Goh, C. Chen, W. Li, Y. 2016 (July). Self-organizing tool for smart design with predictive customer needs and wants to realize Industry 4.0. *In: Proceedings of the 2016 IEEE*

*Congress on Evolutionary Computation (CEC)*, Vancouver, BC, Canada, 24–29 July. IEEE: Piscataway, NJ, USA.

Scholl, H.J. and Al-Awadhi, S. 2016. Creating Smart Governance: The key to radical ICT overhaul at the City of Munich. *Information Polity, 21*(1): 21–42.

Schumacher, A., Erol, S. and Sihn, W. 2016. A maturity model for assessing industry 4.0 readiness and maturity of manufacturing enterprises. *Procedia CIRP, 52*: 161–166.

Schwab, K. 2016. *The Fourth Industrial Revolution*. Davos: World Economic Forum.

Schwab, K. 2016. *The Fourth Industrial Revolution: What It Means, How to Respond.* World Economic Forum. Available at: https://www.weforum.org/agenda/2016/01/the-fourth-industrial-revolution-what-it-means-and-how-torespond/.

Schwab, K. 2018. The Fourth Industrial Revolution. *Encyclopaedia Britannica*. https://www.britannica.com/topic/The-Fourth-Industrial-Revolution-2119734.

Senn, C. 2019. *The Nine Pillars of Industry 4.0*. [Online:] https://www.idashboards.com/ blog/2019/07/31/the-pillars-of-industry-4-0/ (Access data: 2019-02-07).

Shohin, A.E. and Runyang, Z. 2018. *IoT-enabled Personalisation for Smart Products and Services in the Context of Industry 4.0*. http://hdl.handle.net/2292/48057.

Siemens, 2017. *Od Industry 4.0 to Smart Factory*. https://publikacje.siemens-info.com/pdf/76/Od%20Industry%204.0%20do%20Smart%20Factory.pdf. (*Poradnik menadżera i inżyniera*). Warszawa, listopad, 2017).

Sniderman, B., Mahto, M. and Cotteleer, M. 2016. *Industry 4.0 and Manufacturing Ecosystems*. Deloitte University Press. https://www2.deloitte.com/content/dam/insights/us/articles/manufacturing-ecosystems-exploring-world-connected-enterprises/DUP_2898_ Industry4.0ManufacturingEco systems.pdf. (Access data: 20.08.2018).

Sodhi, H. 2020. When Industry 4.0 meets Lean Six Sigma: A review. *Ind. Eng. J., 3*: 1–12.

Suoniemi, S., Meyer-Waarden, L., Munzel, A., Zablah, A.R. and Straub, D. 2020. Big data and firm performance: The roles of market-directed capabilities and business strategy. *Inf. Manag., 57*: 103365.

Ślusarczyk, B. 2018. Industry 4.0—Are We Ready? *Pol. J. Manag. Stud., 17*: 232–248

Taleb, N.N. 2008. *The Black Swarm*. Penguin Books Ltd. (UK).

Thoben, K.D., Wiesner, S. and Wuest, T. 2017. "Industrie 4.0" and smart manufacturing: A review of research issues and application examples. *International Journal of Automation Technology, 11*(1): 4–16.

Tiwari, S. 2021. Supply chain integration and Industry 4.0: A systematic literature review. *Benchmarking Int. J., 28*: 990–1030.

Tiwari, A. and Jain, M. 2013. Analysis of supply chain management in cloud computing. *Int. J. Innov. Technol. Explor. Eng., 3*: 152–155.

United Nations, 2016. *United Nations e-government Global Survey*. Retrieved from https://publicadministration.un.org/egovkb/en-us/Reports/UN-E-Government-Survey-2016.

Ustundag, A. and Cevikcan E. 2018 (January). Industry 4.0: Managing The Digital Transformation. Springer. DOI: 10.1007/978-3-319-57870-5.

Wang, G., Gunasekaran, A., Ngai, E.W. and Papadopoulos, T., 2016. Big data analytics in logistics and supply chain management: Certain investigations for research and applications. *Int. J. Prod. Econ., 176*: 98–110.

Waidner, M. and Kasper, M. 2016. Security in Industrie 4.0:- Challenges and Solutions for the Fourth Industrial Revolution. *EDAA, 2016*: 1303–1308.

Waschull, S., Bokhorst, J.A., Molleman, E. and Wortmann, J.C. 2020. Work design in future industrial production: Transforming towards cyber-physical systems. *Comput. Ind. Eng., 139*: 105679.

Weng, W.H. 2020. Internet of things utilization in marketing for competitive advantage: An organizational capability perspective. *In: Proceedings of the 20th International Conference on Electronic Business, ICEB'20*, 200–209. Hong Kong SAR, China, 5–8 December.

Weyer, S., Schmitt, M., Ohmer, M. and Gorecky, D. 2015. Towards Industry 4.0-Standardization as the crucial challenge for highly modular, multi-vendor production systems. *IFAC-Pap., 48*: 579–584.

Wrycza, S. and Maślankowski, J. (Eds.). 2019. *Informatyka ekonomiczna: Teoria i zastosowania.* Wydawnictwo Naukowe PWN, Warsaw.

World Economic Forum, 2016. *The Future of Jobs: Employment, Skills, and Workforce Strategy for the Fourth Industrial Revolution.* World Economic Forum, Geneva, Switzerland.

Xu, L.D. and Duan, L. 2019. Big data for cyber-physical systems in industry 4.0: A survey. *Enterp. Inf. Syst., 13*: 148–169.

Xu, X., Lu, Y.Q., (...), Wang, L.H. 2021. Industry 4.0 and Industry 5.0 – Inception, conception, and perception. *Journal of Manufacturing Systems, 61*: 530–535.

www.plattform-i40.de.

Yaqub, M.Z. and Alsabban, A. 2023. Industry-4.0-Enabled Digital Transformation: Prospects, Instruments, Challenges, and Implications for Business Strategies. *Sustainability, 15*: 8553. https://doi.org/10.3390/su15118553.

Zhan, Y., Tan, K.H., Li, Y. and Tse, Y.K. 2018. Unlocking the power of big data in new product development. *Ann. Oper. Res., 270*: 577–595.

Zhong, R.Y., Dai, Q.Y., Qu, T., Hu, G.J. and Huang, G.Q. 2013. RFID-enabled real-time manufacturing execution system for mass customization production. *Robot. Comput. Integr. Manuf., 29*: 283–292.

Zhong, R.Y., Xu, X., Klotz, E. and Newman, S.T. 2017. Intelligent manufacturing in the context of industry 4.0: A review. *Engineering, 3*: 616–630.

Zhou, K., Liu, T. and Zhou, L. 2015 (August). Industry 4.0: Towards future industrial opportunities and challenges. *In: 2015 12th International Conference on Fuzzy Systems and Knowledge Discovery (FSKD),* 2147–2152.

# 2 | Business Maturity Models

This chapter examines the conceptual approach to business maturity models, emphasizing their dynamic nature and association with organizational change and development. The text explores Maturity Models (MMs) as structured sequences applicable to diverse entities, elucidating the expected developmental trajectory. Business process maturity, rooted in concepts like Total Quality Management and Business Process Management, is scrutinized as an indicator of an organization's optimization and effectiveness. The chapter delves into the multifaceted application of MMs, including digital transformation and organizational process maturity, discussing benefits and challenges associated with Business Maturity Models (BMMs). Despite criticisms, the text underscores the practical utility of these models as valuable tools for a process-oriented approach, advocating for empirical methodologies and stakeholder involvement in their development. Furthermore, the chapter delves into organizational process maturity, quality management systems, and technological readiness, highlighting models such as PEMM™ and ISO standards. It concludes with an in-depth exploration of Industry 4.0 maturity models, emphasizing frameworks like SIMMI 4.0 and MEMM, and addressing the significance of strategies, collaboration, and leadership in navigating the challenges of the fourth industrial revolution.

## 1.  Conceptual Approach to Business Maturity Model Designation

Maturity is the ability to change and develop. This concept pertains to phenomena and processes associated with improving skills and acquiring certain traits. It signifies readiness to fulfill specific tasks. Manifestations of maturity include cited qualities such as efficiency, effectiveness, and excellence. In management sciences, organizational maturity is understood as a certain level of skills, including excellence, and the degree of an organization's readiness to perform tasks and achieve goals.

Maturity generally signifies a certain advancement in the evolution of a system toward a predefined state (Schumacher et al., 2016). It can be viewed as an incremental process characterized by multiple stages arranged in a specific order, with each stage delineating the criteria for a particular level of complexity. Consequently, a Maturity Model (MM) encompasses a structured sequence of maturity levels applicable to a category of entities, outlining the anticipated, desired, or typical developmental trajectory of these entities across successive discrete stages. The initial level signifies a starting point, while the highest level signifies complete maturity within the specified domain (Becker et al., 2009). Utilizing presumed patterns, an MM can illustrate how the capabilities of an organization progressively develop along an expected, desired, or logical path of maturity. The extent of maturity can be assessed either qualitatively or quantitatively, using either discrete or continuous metrics (Pöppelbuß and Röglinger, 2011).

The utilization of MMs can be facilitated through pre-established methodologies, such as the use of surveys or questionnaires. Following the assessment of the current state (as-is analysis), suggestions for enhancement measures can be drawn and organized in order to advance to higher maturity levels (Hein-Pensel et al., 2023). MMs are customarily crafted to cater to particular use cases, domains, and industries (Fukas et al., 2021). Within the realm of digital transformation, employing an MM can serve as a valuable roadmap, pinpointing and addressing disparities between the existing state and the desired target state (Limat, 2022).

Business process maturity refers to the level of optimization, standardization, and effectiveness of an organization's business processes. It is a measure of how well an organization manages and executes its various processes to achieve its goals and objectives. The concept is often associated with Business Process Management (BPM) and is used to assess the organization's ability to consistently deliver quality products or services while continuously improving its processes (Hyrynsalami et al., 2023).

The methods for assessing and measuring the maturity level in quality management systems stem from models of organizational process maturity. The concept of process maturity, in turn, originates from the ideas of Total Quality Management (TQM) and BPM. Business Process Maturity Models (BPMM) are sets of recommendations and best practices for achieving operational efficiency in implemented processes.

The notion of maturity was first proposed by P. Crosby, who defined it as a state of completeness, perfection, and readiness. Organizational process maturity can be defined as the extent to which processes are formally defined, managed, flexible, measured, and effective. Process maturity is a concept inspired by both quality management and the principles of good business practices (Depaoli, 2013).

In this context, an organization that is mature in terms of processes can be understood as one whose processes are considered mature from a qualitative perspective. Maturity in the context of processes can also be defined as the ability of the organization and its implemented processes to systematically deliver increasingly better results in its activities.

In a functional approach to organizational process maturity, the degree of advancement in the applied methods and techniques of process management, as well as the level of awareness and knowledge about the functioning of processes in the organization, play a decisive role in determining process maturity. This knowledge is used for decision-making by the leadership (Surkat and Leeraphong, 2024).

An MM is a conceptual framework that provides a structured and phased approach to assess and improve the capabilities and performance of an organization or system in a specific discipline or domain. It serves as a roadmap for organizations to understand their current state, set goals for improvement, and systematically advance through different stages of maturity.

In the case of assessing process maturity, it is necessary to use an existing model or develop a custom process maturity model. The assessment of process maturity can be carried out according to the following scheme (Glaveli et al., 2023; Skyttermoen and Wedum, 2023):

- Selection of a process maturity model (or development of a new one),
- Parameterization of the model, taking into account industry specificity,
- Collection of empirical data,
- Data analysis,
- Preparation of the final report.

Business Maturity Models (BMMs) provide a structured framework for organizations to evaluate their current state, identify areas for improvement, and establish a roadmap for future development. These models consider various dimensions of an organization, encompassing processes, technology, culture, and leadership, among others. By offering a holistic view, BMMs help businesses to align their strategies with their overall maturity level, fostering resilience and adaptability in the face of ongoing transformations (Alabbadi et al., 2023).

The concept of MMs is rooted in the idea that organizations evolve and improve over time in a predictable and measurable manner. These models typically consist of multiple levels or stages, each representing a degree of maturity in the targeted area. As an organization progresses through these levels, it demonstrates an increasing ability to manage processes, resources, and quality effectively within that discipline.

MMs are applicable across various fields, including project management, software development, data management, and organizational processes. They are designed to be adaptable to the specific characteristics and needs of the domain they address. The assessment criteria often include qualitative and quantitative measures, focusing on aspects such as people, culture, processes, structures, technology, and other relevant factors (Oruthotaarachchi and Wijayanayake, 2023).

The ultimate goal of using MMs is to help organizations enhance their capabilities systematically, identify areas for improvement, and establish a benchmark for ongoing development. This structured approach allows organizations to make informed decisions, allocate resources efficiently, and continuously strive for excellence in the targeted discipline.

MMs serve as streamlined depictions of the potential for ongoing enhancement within a specific field. These models evaluate the effectiveness of your organization or system in advancing from a defined starting point, providing an assessment of the company's maturity within the realms of quality or resources associated with the discipline. Typically, MMs take into account qualitative data when evaluating aspects such as individuals, cultural factors, processes, organizational structures, physical entities, and technological components (Maris et al., 2023).

For example, within a data technology MM, we can pinpoint various levels of maturity pertaining to a company's utilization of data. Each level within the maturity model delineates the progression of a business as it incorporates data in a particular manner.

In the Table 1 there is an overview of main features of maturity business models. The description emphasizes their purpose, structure, assessment methodology, and the guidance they provide for organizational improvement.

**Table 1**  Main features of business maturity models.

| *Feature* | *Description* |
| --- | --- |
| **Purpose/Objective** | Clearly defines the purpose and objectives of the maturity model, such as improving organizational performance, enhancing processes, or achieving specific business goals. |
| **Stages/Levels** | Typically consists of distinct stages or levels that represent different levels of organizational maturity. Each stage signifies a higher level of capability, efficiency, and effectiveness in managing business processes. |
| **Criteria/ Attributes** | Specifies the criteria or attributes used to assess maturity at each level. These may include organizational culture, leadership, processes, technology, and other relevant factors depending on the specific focus of the maturity model. |
| **Assessment Methodology** | Defines the approach or methodology for assessing the organization's maturity level. This may involve self-assessment, external audits, surveys, interviews, or a combination of these methods. The assessment process should be systematic and repeatable. |
| **Key Performance Indicators (KPIs)** | Identifies key performance indicators that measure the organization's performance at each maturity level. These metrics help track progress and provide a quantitative basis for assessing maturity and improvement efforts. |
| **Roadmap/ Guidance for Improvement** | Offers a roadmap or guidance for organizations to progress from one maturity level to the next. This may include recommended actions, best practices, and strategies for improvement, helping organizations navigate their journey towards maturity. |
| **Integration with Business Strategy** | Aligns with and supports the overall business strategy and objectives. The maturity model should be integrated into the organization's strategic planning process, ensuring that improvements contribute to the achievement of broader business goals. |

*Contd.*

**Table 1** *Contd.*

| Feature | Description |
|---|---|
| **Flexibility/ Adaptability** | Recognizes that different organizations may have unique contexts and requirements. A good maturity model should be flexible and adaptable, allowing organizations to tailor their approach based on their specific industry, size, and operational nuances. |
| **Communication/ Change Management** | Emphasizes the importance of effective communication and change management throughout the maturity improvement process. Clear communication ensures that stakeholders understand the purpose, benefits, and expectations associated with the maturity model. |
| **Continuous Monitoring and Review** | Highlights the need for ongoing monitoring and regular reviews of maturity levels. This ensures that the organization remains dynamic and responsive to changes in its internal and external environment, enabling continuous improvement over time. |
| **Benchmarking and Best Practices** | Encourages benchmarking against industry standards and best practices. Organizations can benefit from comparing their maturity levels to those of industry leaders, identifying areas for improvement and adopting proven strategies for enhanced performance. |

*Source:* Own elaboration based on (Hyrynsalami et al., 2023; Glaveli et al., 2023; Skyttermoen and Wedum, 2023; Maris et al., 2023; Oruthotaarachchi and Wijayanayake, 2023; Ehrensperger et al., 2023).

The business process maturity model employs a five-tier framework to evaluate the maturity of an organization (Table 2). These levels include (Alabbadi et al., 2023; Glaveli et al., 2023; Skyttermoen and Wedum, 2023; Maris et al., 2023; Maris et al., 2023; Kozlov et al., 2023):

- **Initial:** The initial stage represents the lowest level, indicating inconsistent management practices or teams that respond to crises rather than anticipating them.
- **Managed:** The second tier describes organizations with a foundational management structure, yet individual teams within the business operate in isolation with minimal collaboration and limited incorporation of improvement strategies.
- **Standardized:** The standardized, or process management, level signifies that the business is conscious of its processes and is actively working toward achieving consistency and uniform delivery.
- **Predictable:** At the predictable level, organizations leverage their process infrastructure and asset capabilities to attain reliable results by controlling variations within their outputs.
- **Optimizing:** At the optimizing level, companies are engaged in continuous improvement efforts and are dedicated to fostering innovation.

Table 2  Typical levels used in maturity models.

| Level | Description |
|---|---|
| **Level 1: Initial** | • Ad-hoc processes and practices with minimal structure.<br>• Lack of standardized procedures and documentation.<br>• Reactive problem-solving rather than proactive planning.<br>• Limited awareness of process improvement opportunities. |
| **Level 2: Repeatable** | • Basic processes are established and documented.<br>• Some standardization in procedures.<br>• Initial focus on process consistency and reliability.<br>• Limited optimization and formalization of processes. |
| **Level 3: Defined** | • Processes are well-defined and documented across the organization.<br>• Standardized procedures are consistently followed.<br>• Greater emphasis on proactive process management.<br>• Ongoing monitoring and measurement of key processes. |
| **Level 4: Managed** | • Quantitative data is used for performance measurement and management.<br>• Process performance is continually monitored and analyzed.<br>• Focus on process optimization and efficiency.<br>• Proactive identification of improvement opportunities. |
| **Level 5: Optimized** | • Continuous improvement is ingrained in the organizational culture.<br>• Processes are continually optimized for efficiency and effectiveness.<br>• Emphasis on innovation and leveraging best practices.<br>• Proactive anticipation of future challenges. |

*Source:* Own elaboration based on: (Alabbadi et al., 2023; Glaveli et al., 2023; Skyttermoen and Wedum, 2023; Maris et al., 2023; Maris et al., 2023; Kozlov et al., 2023).

The levels of process maturity signify the progressive development of a process. Typically, process MMs encompass three to seven levels, each delineating the expected characteristics a company or process should exhibit before advancing to the subsequent stage.

The rate of maturation varies across processes, influenced by their inherent nature and the priority a business assigns to process enhancement. Nurturing skills and capabilities at each maturity level establishes a foundation for success. Inconsistencies in maturity levels may exist among processes within a company or individual departments. Without vigilant monitoring and a commitment to continuous improvement, processes may regress to lower maturity levels.

Attaining the highest maturity levels is not a universal necessity for all organizations or processes. While high maturity in essential processes may align with strategic objectives, less critical processes might adequately function at lower maturity levels. Notably, maturity models do not prescribe a specific maturity level for an organization or process. According to Van Looy, determining the optimal level is pivotal, and maturity models merely describe capability areas, leaving the decision of the optimal level to the organization based on its strategy (Van Looy, 2018).

The models for assessing process maturity should primarily be seen as tools that enable managers to describe and analyze the current state of the organization (as-is state) and establish the desired future state (to-be state). Furthermore, by applying maturity assessment models, it is possible to identify weaknesses in the implemented processes, as well as within the entire organization, and focus on their elimination (process improvement). It is worth emphasizing that improving processes using these models can be relatively easier due to the inclusion of best practices and recommendations that define the best ways to achieve the desired state and avoid associated problems (de Bruin et al., 2005).

Process maturity assessment models, depending on the organization's needs, can be practically applied in various situations. As mentioned earlier, their fundamental assumption is the identification of the current level of maturity of implemented processes and determining ways to achieve a higher level of maturity. However, depending on the organization's specifics and requirements set by, among others, managers responsible for process management, maturity assessment models can serve various functions (Becker, 2009):

- **Descriptive:** These models are used for the ongoing assessment of implemented processes, considering the criteria within them (as-is assessment). Such an assessment can be internal (conducted independently by the organization) or external (performed by an independent unit), and its results can be communicated to internal or external stakeholders.
- **Prescriptive** (also translated as arbitrary): These models help identify the target level of process maturity (to-be assessment) and provide recommendations for necessary improvement actions.
- **Comparative:** These models serve as a reference model for internal or external comparisons. With historical data on process performance, it is possible to compare their maturity across individual organizational units or entire enterprises (essentially similar to benchmarking).

Achieving a higher level of business process maturity is generally associated with increased efficiency, reduced costs, improved quality, and enhanced customer satisfaction. Organizations often strive to move up the maturity levels to remain competitive and better respond to dynamic business environments. The journey towards higher maturity involves a combination of process optimization, technology adoption, employee training, and a commitment to a culture of continuous improvement (Merdin et al., 2023).

In the Table 3, there is an overview of main benefits connected with usage of BMMs in organizations.

**Table 3**   Benefits of business maturity model usage in organizations.

| *Benefit* | *Description* |
|---|---|
| Strategic Alignment | Business maturity models help organizations align their strategies with their current capabilities and future goals, ensuring that efforts are focused on areas that drive long-term success. |

*Contd.*

**Table 3** *Contd.*

| Benefit | Description |
|---|---|
| Performance Assessment | These models provide a systematic framework for assessing and measuring the performance of various business functions, allowing organizations to identify strengths and weaknesses. |
| Continuous Improvement | By offering a structured approach to evaluating maturity levels, organizations can identify opportunities for improvement and implement a continuous improvement culture for sustained growth. |
| Risk Mitigation | Maturity models enable organizations to identify and mitigate risks associated with different business processes, reducing the likelihood of disruptions and enhancing overall resilience. |
| Resource Optimization | Through the identification of maturity levels, organizations can optimize resource allocation by directing investments and efforts toward areas that will have the most significant impact on overall performance. |
| Benchmarking | Maturity models provide a basis for benchmarking against industry standards and best practices, allowing organizations to compare their maturity levels with peers and competitors for informed decision-making. |
| Enhanced Decision-Making | Organizations can make more informed and strategic decisions by leveraging insights gained from maturity assessments, leading to better allocation of resources and improved overall decision-making processes. |
| Stakeholder Communication | Maturity models facilitate clear communication with stakeholders by providing a standardized language and framework for discussing business processes, performance, and improvement initiatives. |
| Change Management | Organizations can use maturity models as tools for effective change management, guiding the implementation of changes in a structured manner and ensuring that improvements align with overall business objectives. |
| Customer Satisfaction | By focusing on maturity in customer-facing processes, organizations can enhance customer satisfaction by delivering consistent and high-quality products or services, meeting or exceeding customer expectations. |

*Source:* Author's own elaboration.

In the Table 4, there is a description of the main problem connected with BMMs' usage in organizations.

**Table 4** Problems of business maturity model usage in organizations.

| Problem | Description |
|---|---|
| Lack of Customization | Business maturity models often provide a standardized framework, which may not align perfectly with the unique characteristics and needs of each organization. This lack of customization can lead to misalignment and the inability to address specific challenges faced by a particular business. |

*Contd.*

**Table 4** *Contd.*

| | |
|---|---|
| Overemphasis on Process | Some maturity models tend to overly focus on process improvement, neglecting other crucial aspects such as organizational culture, innovation, and adaptability. This narrow focus may hinder overall organizational growth and resilience in dynamic business environments. |
| Rigidity in Implementation | Organizations may face challenges in implementing a business maturity model due to its rigid structure. The rigid nature of some models can result in resistance from employees, making it difficult to adapt the model to the organization's evolving needs and changing circumstances. |
| Inaccurate Self-Assessment | Organizations may struggle with accurately assessing their own maturity level. Bias, lack of objectivity, or a limited understanding of the model can lead to inaccurate self-assessment, preventing the organization from identifying and addressing its actual strengths and weaknesses. |
| Time and Resource Intensive | Implementing and maintaining a business maturity model can be resource-intensive, requiring substantial time, financial investment, and personnel commitment. This may pose challenges, particularly for smaller organizations with limited resources or those facing urgent operational issues. |
| Short-Term Focus, Neglecting Long-Term Goals | Some maturity models may encourage organizations to prioritize short-term gains at the expense of long-term strategic goals. This can result in a myopic focus on immediate improvements without considering the sustainability and long-term success of the organization. |
| Resistance to Change | Employees may resist the changes associated with implementing a maturity model. Resistance can stem from fear of job displacement, unfamiliarity with new processes, or a general aversion to change. Overcoming this resistance is crucial for successful adoption and implementation. |
| Lack of Alignment with Industry Dynamics | Business maturity models may not always align with the rapidly changing dynamics of specific industries. A model that fails to account for industry-specific trends and challenges may provide limited value and hinder an organization's ability to stay competitive in its market. |
| Insufficient Communication and Training | Successful implementation requires effective communication and training. Inadequate communication about the purpose and benefits of the maturity model, along with insufficient training on how to use it, can impede understanding and hinder the organization's ability to derive value from the model. |
| Tendency to Stifle Innovation | Some maturity models, by emphasizing standardization and conformity, may unintentionally stifle innovation within an organization. This can be detrimental, especially in industries where rapid innovation is essential for staying ahead of the competition and adapting to market changes. |

*Source:* Own elaboration based on: (Merdin et al., 2023; Das et al., 2023; Alabbadi et al., 2023; Glaveli et al., 2023; Skyttermoen and Wedum, 2023).

The Process Maturity can be divided into two groups: Low-Level Process Maturity and High-Level Process Maturity. Low-Level Process Maturity is characterized by limited leadership involvement and guidance, a resistance to change, incomplete or inconsistent process documentation, limited training programs and resources, unclear or fragmented ownership of processes, a lack of metrics and performance measures, a reactive approach to problem-solving, outdated or underutilized technology, limited awareness and mitigation strategies for risks, inconsistent customer satisfaction, and challenges in adaptability (Ahmad et al., 2024).

High-Level Process Maturity is marked by strong, proactive leadership driving initiatives, a culture of continuous improvement and innovation, comprehensive, up-to-date documentation and standards, ongoing, targeted training to enhance skills and knowledge, clearly defined ownership and accountability for processes, a robust metrics framework with continuous monitoring, a proactive and systematic approach to continuous improvement, advanced and integrated technology supporting processes, proactive risk identification and mitigation planning, a strong customer-centric focus with tailored solutions, and agility and adaptability to evolving business needs (Tripathi et al., 2024; Ouazani-Chahidi et al., 2024). Table 5 shows a comparison between Low-Level and High-Level Process Maturity.

**Table 5** Comparison between Low-Level and High-Level Process Maturity.

| Aspect | Low-Level Process Maturity | High-Level Process Maturity |
| --- | --- | --- |
| Leadership | Limited leadership involvement and guidance | Strong, proactive leadership driving initiatives |
| Culture | Resistance to change, lack of process awareness | A culture of continuous improvement and innovation |
| Documentation | Incomplete or inconsistent process documentation | Comprehensive, up-to-date documentation and standards |
| Training | Limited training programs and resources | Ongoing, targeted training to enhance skills and knowledge |
| Process Ownership | Unclear or fragmented ownership of processes | Clearly defined ownership and accountability for processes |
| Metrics and Measurement | Limited use of metrics and performance measures | Robust metrics framework with continuous monitoring |
| Continuous Improvement | Reactive approach to problem-solving | Proactive, systematic approach to continuous improvement |
| Technology Integration | Outdated or underutilized technology | Advanced, integrated technology supporting processes |
| Risk Management | Limited awareness and mitigation strategies | Proactive risk identification and mitigation planning |
| Customer Focus | Inconsistent customer satisfaction | Strong customer-centric focus with tailored solutions |
| Adaptability | Resistance to change and adaptability challenges | Agile and adaptable to evolving business needs |

*Source:* Author's own elaboration on basis: (Eby, 2022; Van Loy, 2014; Tripathi et al., 2024; Ouazani-Chahidi et al., 2024).

The models offer a simplified representation of reality, aiding in the comprehension of phenomena associated with processes. However, they do not claim to encompass all potential situations. The model provides a direction and guidance without enforcing specific solutions (Jesus et al., 2022). The primary criticism of these models lies in their derivation from the consensus of knowledge among specialists based on best practices, resulting in a lack of clear, measurable criteria or indicators for individual levels, and hindering precise determination of their positions. In the business realm, companies typically exist between the described phases rather than conforming strictly to the presented models (Devi et al., 2022).

Further criticism is directed at the mechanistic nature of model construction, contrasting with the complexity of organizations. The success of implementing changes predominantly relies on managers and their attitudes (Richter et al., 2020). Despite the apparent universality of the presented models, it's important to note that they are often developed based on the experiences of large entities, and the extent to which they can be effectively applied in other contexts is insufficiently researched and described (Śliż, 2022; Correia, 2021).

Notwithstanding the criticism, the models discussed in the study prove to be valuable tools for implementing a process-oriented approach. They function as maps for actions, initiatives, or projects aimed at introducing changes in processes (Dewi and Mahendrawathi, 2019). Serving both as tools for retrospective description and for defining essential milestones in achieving higher maturity levels from a prospective standpoint, the models offer practical utility. Given their substantial similarities and shared elements, choosing a model for practical applications is considered a low-risk decision (Glavan, 2019).

Critiques of current MMs frequently highlight the absence of empirical methodologies in their formulation, leading to the replication of model structures based on theoretical constructs. To enhance acceptance and guarantee the applicability of outcomes, a crucial factor necissitate involving industry partners and key stakeholders in the early stages of MM development (Hein-Pensel et al., 2023).

## 2. Review of Business Maturity Models in Modern Industrial Transformation

Modern industrial transformation involves a multifaceted and dynamic process that integrates advanced technologies to enhance productivity, efficiency, and adaptability within industrial sectors. This transformation encompasses a range of interconnected elements, starting with the integration of digital technologies into traditional manufacturing processes (Alshammari, 2023).

The foundation of modern industrial transformation lies in the adoption of Industry 4.0 principles, which emphasize the use of interconnected systems, the Internet of Things (IoT), and Data Analytics. This connectivity enables seamless communication between machines, devices, and systems, fostering a more responsive and intelligent manufacturing environment (Liu et al., 2023).

Central to this transformation is the deployment of smart manufacturing technologies, such as automation, robotics, and Artificial Intelligence (AI). These technologies contribute to the creation of smart factories where machines can make data-driven decisions and adapt to changing production requirements in real time. Automation not only streamlines routine tasks, but also enhances precision and reduces the likelihood of errors (Oruthotaarachchi and Wijayanayake, 2023).

In addition, data plays an important role in the modern industrial landscape. The collection, analysis, and utilization of data generated throughout the production process empower companies to make informed decisions, optimize workflows, and predict maintenance needs (El Chaabi et al., 2023). Big Data Analytics, Machine Learning (ML), and predictive maintenance algorithms enable proactive problem-solving and contribute to overall operational excellence (Khraiwesh, 2020).

Also, human-machine collaboration is a key aspect of modern industrial transformation. Augmented Reality (AR) and Virtual Reality (VR) technologies are integrated into the manufacturing process to provide workers with real-time information, guidance, and training. This facilitates a more flexible and adaptive workforce capable of handling complex tasks and responding to changing production demands (Wood and Vickers, 2018).

Cybersecurity is a critical consideration in modern industrial transformation, as the increased connectivity and reliance on digital technologies expose industrial systems to potential cyber threats. Implementing robust cybersecurity measures is essential to safeguard sensitive data, maintain the integrity of manufacturing processes, and ensure the resilience of the overall industrial ecosystem (Gao et al., 2023).

MMs serve as invaluable tools in the context of modern industrial transformation, offering a structured framework for organizations to assess and enhance their capabilities in adapting to evolving technological landscapes (Chockalingam et al., 2023). These models provide a systematic way to evaluate the maturity of various aspects within an organization, including processes, technologies, and overall readiness for transformation (Glaveli et al., 2023).

One fundamental reason for the use of MMs in industrial transformation is the complexity inherent in the adoption of advanced technologies. By offering a structured and phased approach, these models guide organizations through a series of defined stages, allowing them to progress gradually and incrementally. This step-by-step progression helps mitigate risks associated with rapid and uncontrolled transformation, ensuring that each stage is thoroughly understood and implemented before moving on to the next (Brand et al., 2023).

MMs also facilitate benchmarking, allowing organizations to compare their progress against established standards or industry best practices. This benchmarking process provides valuable insights into areas where improvement is needed, helping organizations identify gaps and prioritize initiatives. Additionally, it enables the sharing of knowledge and experiences within the industry, fostering collaboration and the development of common standards (Maris et al., 2023).

It can be stated that MMs contribute to the establishment of a clear roadmap for industrial transformation. By delineating stages of maturity, these models assist organizations in setting realistic goals, defining achievable milestones, and aligning transformation efforts with overall business objectives. This strategic alignment ensures that the transformation process is not only technologically driven but also aligned with the organization's overarching goals and vision (Wang et al., 2022).

Also, MMs promote a culture of continuous improvement. They encourage organizations to regularly reassess their maturity levels, identify opportunities for optimization, and implement ongoing enhancements. This iterative approach fosters agility and adaptability, allowing organizations to stay ahead of technological advancements and remain competitive in the rapidly evolving industrial landscape (Merdin et al., 2023).

**The Crosby Maturity Grid**

Crosby published his MM in 1979 in the book *Quality is Free.* The maturity grid system is considered the first MM. It focused on quality management and provided a description of five levels of organizational proficiency in using quality management methods and tools. Many subsequent MMs were later inspired by this five-level concept (Crosby, 1979).

Using a questionnaire based on this five-level scale, organizations could assess their proficiency in utilizing quality management tools and techniques, ranging from 1 to 5. Additionally, the model outlined the development path for these skills, specifying the actions that need to be implemented to achieve the next level of maturity.

The Crosby Maturity Grid is a model developed by Philip Crosby, a quality management expert, to assess the maturity of an organization's quality management processes. This grid is designed to help organizations understand their progress in implementing and managing quality improvement initiatives. The model consists of five levels, each representing a stage in the organization's maturity regarding quality management (Sansalvador and Brotons, 2017).

At the initial stage, Level 1, organizations are characterized by a lack of awareness and commitment to quality improvement. As they move to Level 2, they begin recognizing the importance of quality and initiate some improvement efforts. Level 3 signifies a more systematic approach, where processes are established and defined, and employees receive training to understand and contribute to quality improvement.

Moving to Level 4, organizations demonstrate a proactive approach to quality, with a focus on prevention rather than correction. Continuous improvement becomes ingrained in the organizational culture. Finally, at Level 5, organizations achieve a state of optimization, where quality is not just a set of processes but an integral part of the organizational DNA, and a commitment to excellence is ingrained at every level (Sansalvador and Brotons, 2023).

The Crosby Maturity Grid serves as a guide for organizations to assess their maturity in quality management and provides a roadmap for advancing through the different levels toward optimal quality performance (Crosby, 1979).

**The Harrington Model of Maturity**

*The Harrington Model:* The model developed by H. Harrington in 1991 can already be considered classical (Gałuszka, 2011). It served as a counterbalance to business process reengineering, advocating for an evolutionary approach. The foundation of the model consists of six successive maturity levels (Harrington, 2006; Kalinowski, 2011):

- **Unknown:** The status of the process has not been determined.
- **Understandable:** It is clear to participants and operates according to established procedures and documentation.
- **Effective:** Processes are systematically measured, improvement has begun, customer expectations are met.
- **Efficient:** The process is on the path to significant performance improvement.
- **Error-Free:** The process is optimized, characterized by high efficiency, and is free of errors.
- **World-Class:** It is continually improved and can serve as a benchmark for other enterprises (process benchmarking).

H. Harrington points out that the key element is the selection of processes. He suggests that even 50% of processes can be at the lowest maturity levels without harm to the program of improving process efficiency and, consequently, the organization as a whole (Harrington, 2006).

**Capability Maturity Model (CMM)**

The starting point for developing the model was W. Humphrey's book titled *Managing the Software Process*, in which he first outlined the principles of process maturity assessment. The foundational Capability Maturity Model (CMM) was created in 1987 (Humphrey, 1989).

The primary goal and underlying assumption of the model were to assert that organizations whose managers comprehend the principles of a process-oriented approach and systematically manage them can respond more effectively and quickly to changing customer requirements and organizational-level goals.

The CMM is a framework that assesses and evaluates the maturity of an organization's software development and process improvement practices. Developed by the Software Engineering Institute (SEI) at the Carnegie Mellon University, the CMM provides a structured approach to understanding and improving an organization's ability to consistently produce high-quality software (North and Coetzee, 2022).

CMM consists of five maturity levels, each representing a stage of evolutionary development in an organization's software processes. These levels are Initial, Repeatable, Defined, Managed, and Optimizing. The model emphasizes the importance of moving through these levels sequentially, with each level building upon the achievements of the previous one.

At the Initial Level, organizations have ad-hoc and chaotic processes, often resulting in unpredictable outcomes. The Repeatable Level introduces some discipline, allowing for the establishment of basic project management practices.

The Defined Level formalizes and standardizes processes across the organization, ensuring consistency and repeatability (Jordon et al., 2022).

As organizations progress to the Managed Level, they implement metrics and controls to monitor and manage their processes effectively. The highest level, Optimizing Level focuses on continuous improvement, innovation, and the proactive identification and management of risks.

CMM is not limited to software development; its principles can be applied to various business processes. The model serves as a benchmark for organizations to assess their current capabilities and set goals for improvement, promoting a systematic and structured approach to achieving higher levels of maturity in their processes (Garbin et al., 2022). By adhering to the CMM framework, organizations can enhance their overall performance and increase the reliability and quality of their software products and services (Chen et al., 2022).

Originally, the CMM, an organizational MM, was developed to assess processes related to software development. In its initial version, it comprised a list of so-called best practices divided into sectors known as process areas (e.g., requirements management, project planning, etc.), forming the basis for the organization's maturity assessment method. Maturity levels in this model were evaluated on a scale from 1 (initial state) to 5 (continuous process improvement) when considering individual process areas. Positive opinions about the effectiveness of this approach quickly extended beyond those interested solely in the field of improving the quality of software development processes.

**Capability Maturity Model Integration (CMMI)**

The Capability Maturity Model Integration (CMMI) is a framework that provides organizations with a set of best practices for improving their processes. It is designed to help organizations enhance their ability to develop and maintain high-quality products and services. CMMI is not specific to any particular industry; instead, it can be applied across various domains, including software development, systems engineering, and project management (Alshammari, 2023; Liu et al., 2023).

The model was originally designed to improve the manufacturing processes of software and information systems. Later, its versatility allowed for its application in diagnosing the maturity levels of business organizations. The model emerged from the evolution of the CMI model, aiming to increase its complexity to make it implementable in organizations with diverse profiles (Amarasekara et al., 2022).

This model enables process improvement through two representations: staged and continuous, containing the same content but with different representations. The staged representation encompasses all processes in the organization, determining their maturity level using five maturity levels. It serves as a reference point for process assessment and comparison with other organizations that have achieved a specific maturity level (Sicheng et al., 2022).

The continuous representation focuses on selected process areas and introduces the concept of performance levels, allowing the organization to refine specific processes. The first version of the CMMI model was developed in 2002, followed by versions in 2006 and 2010. The latest version, known as CMMI 2.0, was created in March 2018. The model includes three fundamental versions:

- **CMMI for Development:** Supporting organizations involved in the development of products or services,
- **CMMI for Services:** Supporting organizations engaged in service delivery,
- **CMMI for Acquisition:** Supporting organizations involved in acquiring products and services from external suppliers.

Each of the models is built on so-called process areas. Process areas can be characterized as a set of interrelated best practices. When implemented collectively, they contribute to achieving organizational goals relevant for improvement in a specific area. For example, among the process areas, one can mention integrated project management, measurement and analysis, and risk management.

Individual models are constructed in such a way that they are partially based on universal (common) process areas, of which there are a total of 16 (an example being risk management). They are also partially based on process areas characteristic of specific types of activities (development work, services, deliveries).

Within the process areas, lists of goals and practices are identified. General and specific goals are distinguished, depending on whether they are applied to all process areas (general) or assigned to a specific one (specific). Practices, in turn, can be defined as actions that are important for achieving the above-mentioned goals.

Within each of the models, there are two so-called representations (CMMI representation) from the perspective of which the analyzed processes in the organization can be considered. The continuous representation allows focusing on specific processes (or process areas) that are important in the context of organizational strategic goals or operational risk minimization. By using this representation, it is possible to establish the profile of an organization through an independent analysis of each process area. Since the maturity level is different in each of them, it is possible to identify the strengths and weaknesses of the organization and, thus, determine improvement plans for each process area (Al-Matari et al., 2021).

On the other hand, the staged representation more effectively allows presenting the state of an organization as a whole rather than its maturity in each process area. Therefore, it has greater applicability in developing a standardized approach to process improvement within an organization (establishing strategies for process improvement, goals, and schedules). It can also be used for conducting comparisons (benchmarking) regarding the maturity of processes between organizations. In summary, using the continuous representation involves assessing the maturity and setting improvement goals for individual processes. In the case of the staged representation, the focus is on evaluating the maturity of the organization through the prism of implemented processes (Doss et al., 2021).

The model identifies five maturity levels. At the lowest level, solutions are generated ad-hoc, lacking predictability of results. At the second level, process management occurs within functional units. The third level introduces economies of scale, with standardized processes meeting stakeholder needs. The fourth level involves the organization using statistical methods to anticipate process results. The fifth level is characterized by utilizing innovation to bridge the gap between observed and desired capabilities for achieving strategic goals (Shen et al., 2021).

The CMMI model was originally designed to improve the processes of software development and information systems. Over time, it turned out to be versatile enough to be applied in practice for diagnosing the maturity levels of business processes. The model allows for process improvement through two representations: staged and continuous, both containing the same content but differing in their presentation in the model (Khraiwesh, 2020).

CMMI was developed by the Software Engineering Institute (SEI) at Carnegie Mellon University and has become a widely recognized and adopted framework for process improvement in various industries. It helps organizations establish a disciplined and systematic approach to managing and improving their processes, leading to better performance and increased customer satisfaction (Wood and Vicker, 2018).

Table 6 has a description of the main features of the CMMI model.

**Table 6**   Main features of CMMI model.

| Feature | Description |
| --- | --- |
| Maturity Levels | Represents stages of organizational process improvement. Levels include Initial, Managed, Defined, Quantitatively Managed, and Optimizing, each reflecting a higher degree of process maturity. |
| Process Areas | Defines specific aspects of an organization's processes that need attention. Examples include Requirements Management, Project Planning, Configuration Management, and Continuous Improvement. |
| Representations | CMMI supports two representations: Staged (organizing practices into maturity levels) and Continuous (allows organizations to select and improve specific process areas independently). |
| Appraisal | Organizations can undergo a CMMI appraisal (e.g., SCAMPI) to assess their current maturity level and identify areas for improvement. This process helps in evaluating and enhancing overall capabilities. |
| Adoption | CMMI is adopted by organizations to improve process capabilities, quality, and project management. It provides a roadmap for moving from ad-hoc processes to well-defined, mature, and continuously improving states. |
| SEI (Software Engineering Institute) | CMMI was developed by SEI at Carnegie Mellon University. SEI provides resources, training, and support for organizations implementing CMMI, contributing to its widespread recognition and adoption. |

*Source:* Own elaboration based on: (Riaz, 2017; Pane and Sarno, 2015; Čengija and Kapitan, 2008).

The staged representation encompasses all processes in the organization simultaneously and determines their maturity level using five levels (Riaz, 2017; Pane and Sarno, 2015; Čengija and Kapitan, 2008):

- **Initial:** Processes are chaotic, actions are conducted ad-hoc budgets, and deadlines are regularly exceeded.
- **Managed:** Processes are planned, managed, measured, and controlled based on a plan; project management is implemented and applied.
- **Defined:** Processes are well-characterized and understood, described in standards and procedures.
- **Quantitatively Managed:** Processes are controlled using statistical methods and other quantitative measurement methods.
- **Optimizing:** Processes are further developed and improved (optimized) to achieve business goals. The causes of process variability are analyzed, and appropriate corrective actions are taken. Measurable goals for process improvement are defined and continuously updated based on changing business situations.

Table 7 contains examples of how organizations might exhibit characteristics at different CMMI maturity levels. Table 8 presents the maturity levels of software development company example.

**Table 7**  Maturity levels in CMMI model.

| Maturity Level | Focus | Example |
|---|---|---|
| **Level 1: Initial** | Unpredictable processes | Ad-hoc development without defined processes. Lack of standardization. |
| **Level 2: Managed** | Basic project management | Basic project planning, tracking, and oversight. Requirements and processes are defined, but may not be consistently followed. |
| **Level 3: Defined** | Standardized processes | Standardized processes are in place and followed across the organization. Training programs ensure staff understands and follows the defined processes. |
| **Level 4: Quantitatively Managed** | Process measurement and control | Metrics are collected and analyzed to manage and control the process quantitatively. Organizations use statistical methods to understand and improve processes. |
| **Level 5: Optimizing** | Continuous process improvement | Continuous process improvement is ingrained in the organizational culture. Feedback mechanisms and lessons learned are actively used to enhance processes continually. |

*Source:* Own elaboration based on: (Kenneally et al., 2023; Kimita et al., 2022; Wang et al., 2022; Peng et al., 2016).

**Table 8**  Maturity levels in CMMI model on software development company example.

| Maturity Level | Focus | Example |
|---|---|---|
| **Level 1: Initial** | Unpredictable processes | Developers work independently without clear guidelines. Each project follows its own approach, leading to inconsistent results. |
| **Level 2: Managed** | Basic project management | The organization introduces a basic project management framework. Project plans are created, and progress is tracked, but processes may not be well-documented or consistently followed. |
| **Level 3: Defined** | Standardized processes | The company establishes standardized development processes, including requirements gathering, testing, and documentation. Training programs ensure that all team members understand and follow these processes. |
| **Level 4: Quantitatively Managed** | Process measurement and control | Metrics such as defect rates and project performance are systematically collected and analyzed. Statistical methods are used to identify areas for improvement and optimize processes. |
| **Level 5: Optimizing** | Continuous process improvement | The organization actively seeks feedback from projects, conducts regular process reviews, and implements improvements based on lessons learned. Continuous improvement becomes a part of the organizational culture. |

*Source:* Own elaboration based on: (Kenneally et al., 2023; Kimita et al., 2022; Wang et al., 2022; Peng et al., 2016).

CMMI is particularly valuable in industries where quality, reliability, and efficiency are critical, such as software development, aerospace, and healthcare. By adopting CMMI, businesses can not only meet industry standards, but also gain a competitive edge by delivering superior products and services. Overall, CMMI serves as a strategic tool for businesses seeking to optimize their processes, enhance organizational performance, and achieve sustained success in a rapidly evolving business landscape.

The staged representation is, therefore, an excellent reference point for conducting process assessments and comparisons with other organizations that have already achieved a certain level of maturity. It is designed in such a way that it does not require organizations to have knowledge of the level of advancement of existing processes, and consequently, the need to choose those that require improvement.

In organizations that already have knowledge about their processes, know their strengths and weaknesses, and have defined the direction of improvements, the continuous representation can be applied. It focuses on selected areas of the process and introduces the concept of performance levels, allowing the organization to improve specific processes (Hassan and Arshad, 2023).

Using such an approach allows for a precise determination of maturity levels for individual processes according to a specified six-step scale (Wang et al., 2022; Hassan and Arshad, 2023; Saad et al., 2021):

- **Incomplete:** The process does not achieve specific goals.
- **Performed:** The process flow is unstable, quality or cost-related goals are not achieved; the organization has already taken the first steps towards raising the maturity level, but it is difficult to determine to what extent it has brought specific benefits.
- **Managed:** The process is planned, its progress is monitored and controlled.
- **Defined:** At this level, process management is proactive, involving an understanding of the mutual dependencies between individual processes.
- **Quantitatively Managed:** This is a process defined and controlled using statistical methods; additionally, the organization precisely defines goals related to the process flow and conducts accurate measurements.
- **Optimizing:** This is a process that is managed quantitatively and systematically improved. It is worth mentioning that the CMMI model demonstrates practical utility as it is continuously developed through the construction of tools.

The implementation of CMMI 2.0 requires a methodical approach, involving the following key steps (Blue People, 2023):

- **Evaluate the current maturity level of the organization:** Prior to commencing the implementation of CMMI 2.0, the organization should evaluate its existing maturity level using a formal appraisal method, such as the Standard CMMI Appraisal Method for Process Improvement (SCAMPI).
- **Develop the implementation strategy:** Drawing from the assessed current maturity level, the organization should devise a comprehensive plan for implementing CMMI 2.0. This plan should delineate the process areas slated for implementation and establish a timeline for execution.
- **Carry out the implementation of process areas:** The organization should execute the implementation of the identified process areas according to the established roadmap. This may involve creating new processes, modifying existing ones, or adopting industry best practices.
- **Monitor and assess process performance:** Continuous monitoring and measurement of the processes are imperative to ensure adherence and the achievement of desired outcomes.
- **Continuously improve the processes:** Based on the feedback obtained from monitoring and measurement, the organization should iteratively enhance and refine its processes, cultivating a culture of ongoing improvement.

In Table 9, there is an overview of usage of CMM integration in various industries.

**Table 9**  Usage of CMMI model in selected industries.

| Industry | Example |
| --- | --- |
| Software Development | • Establishing standardized software development processes, including requirements analysis, design, coding, testing, and maintenance.<br>• Implementing best practices for code reviews, testing, and release management. |
| Aerospace | • Ensuring compliance with industry standards for safety-critical systems.<br>• Implementing rigorous processes for design verification, validation, and configuration management in aircraft and spacecraft development. |
| Healthcare | • Standardizing processes for medical device development and regulatory compliance.<br>• Ensuring quality and safety in healthcare software development and systems integration. |
| Manufacturing | • Optimizing manufacturing processes to enhance product quality and reduce defects.<br>• Implementing best practices for supply chain management, production planning, and quality control. |
| Defense | • Adhering to strict standards for defense systems' development, including cybersecurity measures.<br>• Implementing risk management and configuration control processes for complex defense projects. |
| Automotive | • Implementing quality management processes in the design, manufacturing, and testing of automotive systems.<br>• Ensuring compliance with industry standards for vehicle safety and emissions control. |
| Telecommunications | • Enhancing the reliability and performance of telecommunications network infrastructure.<br>• Standardizing processes for software development, network design, and system integration. |
| Financial Services | • Implementing robust software development and testing processes for financial applications.<br>• Ensuring compliance with regulatory requirements for data security and privacy in financial transactions. |
| Healthcare IT | • Developing and maintaining Electronic Health Records (EHR) systems with a focus on data integrity and patient privacy.<br>• Ensuring interoperability and seamless integration of healthcare IT systems. |
| Consulting and Services | • Establishing standardized project management processes for consulting engagements.<br>• Continuous improvement in service delivery processes to enhance client satisfaction. |

*Source:* Own elaboration based on: (Wang et al., 2022; Hassan and Arshad, 2023; Saad et al., 2021; Ozkan et al., 2023; Gracey, 2020; Kenneally et al., 2023; Erhard et al., 2023; Zhai et al., 2022).

**Gruchman Model of Process Maturity for Companies**

In this model, four phases are distinguished (Bitkowska, 2009):

- **Zero Phase:** Processes are invisible, in other words, they exist but no one notices them. This paradox is easy to explain: the company in this phase focuses on its organizational structure, essentially functional.
- **Process Initiatives Phase:** Processes are now visible. The same traditional organizational structure still applies, but business process maps emerge, which should encompass all company processes or at least those considered crucial. These maps should reveal horizontal connections and dependencies between activities in different organizationally distinct areas of the company. Thanks to such maps, awareness and understanding of these connections begin to develop in the company.
- **Process Management Phase:** All company processes are inventoried and described. Process descriptions are continuously updated. The word 'process' is on everyone's lips. Awareness and practical knowledge of dependencies within the company are widespread, and, most importantly, individuals responsible for processes, commonly known as their owners or leaders, emerge in this phase.
- **Phase of Process-oriented Organizational Structure:** The institutionalization of processes reaches the highest level. Business processes are at least as important as departments and organizational units.

**Process and Enterprise Maturity Model (PEMM™)**

In 2006, Dr. Michael Hammer from the Harvard Business School introduced the Process and Enterprise Maturity Model, or PEMM™, building on his prior work in process reengineering. Hammer advocated for this model as an objective means to assess business processes. The intention was to create a user-friendly model that organizations could independently apply for assessments, eliminating the need for external experts or consultants (Duan et al., 2023). This approach is valuable because employees are more likely to trust and act upon self-assessments compared to recommendations from external sources. The model can be universally applied within an organization to establish a standardized approach for enhancing process maturity and fostering a common language for comparison (Song et al., 2023; Shmeleva et al., 2023).

The PEMM™ is a comprehensive framework designed to assess and enhance the maturity of both processes and enterprises within an organization. Developed as a strategic tool, PEMM™ provides a structured approach for organizations to evaluate and improve their capabilities across various domains (Eby, 2022).

According to Hammer's 2007 article in the *Harvard Business Review*, successful business processes require two elements: process enablers and organizational capabilities (Hammer, 2007). Process enablers are interconnected individual processes, such as design, performance, process ownership, infrastructure, and metrics, that rely on each other for success. Organizational capabilities refer to the company's ability to create a supportive environment for these process enablers, encompassing leadership, culture, expertise, and governance.

To assess organization-wide process maturity using PEMM™, Hammer provided an evaluating worksheet that measures organizational capabilities on a scale of E0 to E4. Once all capabilities reach a specific level, organizations can align their processes accordingly. Additionally, Hammer's assessing worksheet can be utilized to rank process maturity from P0 to P4. Both worksheets are freely available and can be color-coded to visually represent maturity scores in a more dynamic manner.

At its core, PEMM™ consists of a set of maturity levels that represent the evolutionary stages of an organization's processes and enterprise capabilities. These levels range from an initial, ad-hoc state to a highly mature and optimized state. The model is not limited to specific industries or sectors, making it adaptable and applicable across a broad spectrum of organizations.

PEMM™ encompasses key dimensions such as process management, organizational culture, technology integration, and overall enterprise governance. It recognizes that process maturity is interconnected with the maturity of the broader enterprise. The model emphasizes the need for a holistic approach, encouraging organizations to address not only the technical aspects of their processes, but also the people, culture, and organizational structure.

Organizations can use PEMM™ to identify their current maturity level, set improvement targets, and develop a roadmap for achieving higher levels of maturity. The model provides a common language for discussing process and enterprise maturity, fostering communication and collaboration across different departments and stakeholders.

Moreover, PEMM™ incorporates feedback loops, allowing organizations to continuously monitor and adapt their maturity improvement efforts. This iterative approach aligns with the dynamic nature of business environments and encourages organizations to stay responsive and agile.

**Assessment of the Maturity Level of Quality Management Systems**

The assessment of the maturity level of a quality management system is a fundamental tool used for a comprehensive examination and analysis of the functioning of this system in an organization. It can be defined as a comprehensive, systematic, and regular review of organizational activities according to an adopted model (Slack et al., 2007).

The assessment of the maturity level of a quality management system, if well-conducted, can serve as an excellent diagnostic tool to identify weaknesses in the quality management system and provide recommendations for improvement. However, the problem lies in the fact that most assessment methods either do not evaluate individual criteria or, as is the case with the EFQM model, do not precisely explain the basis for assigning weights to various areas (Tairi, 2010).

In the literature, there are many improvement models used in different philosophies and approaches to quality management. Models used in ISO 9000 series standards, such as ISO 9004:2018, focus on improvement models in organizations. Similarly, quality awards worldwide and in Poland use various improvement models, showing significant similarities (ISO 9004:2018).

The assessment of the maturity level of quality management systems commonly employs various self-assessment models within organizations. It is associated with so-called excellence models and quality awards. It was initially applied on a broader scale and is currently also used in the PN-ISO 10014:2021 standard. Self-assessment is not a new method; it is often used in environmental accreditation procedures for healthcare facilities, accreditation of higher education institutions, and is a qualification condition for companies in successive stages of quality awards. Since many organizations struggle with measuring progress in the systemic approach to quality management or TQM implementation, self-assessment provides convenient tools for a comprehensive assessment of these advancements (ISO 10014:2021).

The discussed assessment technique is very useful for any organization aspiring to develop, especially to improve and monitor the implementation of modern quality management concepts. A systematic review and measurement of organizational performance in its most important aspects should be one of the fundamental activities in a system based on the PN-EN ISO 9001:2015 standard or the implementation of TQM principles (ISO 9001:2015).

The application of methods to assess the maturity level of a quality management system in an organization allows for a precise determination of its strengths and weaknesses, as well as areas requiring improvement. Therefore, it is a very effective tool for continuous improvement. The assessment process should result in the development of planned improvement actions for the organization's functioning and its systematic review and evaluation. Conducting a maturity level assessment allows for recognizing the causes of critical situations and formulating conclusions that serve the organization in making necessary changes in management.

Improvement should be based on a process approach consistent with the PDCA continuous improvement cycle. It is convenient to use the self-assessment model from the PN-ISO 10014 standard in this case, where improvement regarding individual management principles is presented in the form of the PDCA cycle. To ensure the proper progress of improvement activities, the organization should (Tairi, 2010; ISO 9004:2018):

- provide opportunities for people in the organization to participate in improvement activities by empowering them for such actions;
- ensure necessary resources;
- establish recognition and reward systems for improvement;
- continually improve the effectiveness and efficiency of the improvement process.

The PN-EN ISO 9004:2018 standard recommends self-assessment, usually conducted by management, whose results should provide an opinion on the effectiveness and efficiency of the quality management system's operation and its maturity level. However, it is essential to remember that such self-assessment does not substitute for internal or external quality auditing. Still, it provides input data for identifying areas in the organization requiring improvement and can be helpful in setting priorities (ISO 9004:2018).

Table 10 shows a comparison of the three approaches to quality management system maturity: ISO 9004:2021, EFQM Excellence Model, and PN-ISO 10014:1018.

**Table 10** Comparison of basic models for assessing the maturity and excellence of a quality management system.

| Feature | ISO 9004:2018 | EFQM Excellence Model | PN-ISO 10014:2021 |
|---|---|---|---|
| Purpose | Provides a framework for achieving sustained success through continuous improvement. | Assesses an organization's overall performance against nine criteria. | Provides guidance on developing, implementing, and maintaining a quality management system. |
| Structure | Five levels of maturity: initial, repeatable, defined, managed, and optimizing. | Nine criteria: leadership, policy and strategy, people, partnerships and resources, processes, products and services, customer results, society results, key performance results, and innovation and learning. | Nine generic management system elements: leadership and commitment, planning, resource management, implementation and operation, measurement analysis and improvement. |
| Focus | Internal quality management. | Overall organizational performance, including quality, customer satisfaction, people, and society. | Quality management systems in general. |
| Application | Suitable for all organizations, regardless of size or industry. | Suitable for all organizations, but particularly those seeking to achieve excellence. | Suitable for all organizations that have a quality management system or are considering implementing one. |
| Benefits | Improves quality, efficiency, and customer satisfaction. | Enhances overall organizational performance, leading to improved financial results, reputation, and innovation. | Contributes to effective quality management system implementation and maintenance. |

*Contd.*

**Table 10** *Contd.*

| Feature | ISO 9004:2018 | EFQM Excellence Model | PN-ISO 10014:2021 |
|---|---|---|---|
| Continuous Improvement | Emphasizes continuous improvement through a PDCA (Plan-Do-Check-Act) cycle. | Integral part of the model, focusing on learning, innovation, and continuous improvement. | Advocates for continuous improvement and the establishment of a learning organization. |
| Leadership and Governance | Highlights the role of leadership in establishing a quality culture and promoting organizational improvement. | Places strong emphasis on leadership, vision, and values as drivers of excellence. | Acknowledges the importance of leadership in ensuring the effectiveness of the quality management system. |
| Measurement and Evaluation | Encourages the use of performance indicators and data for decision-making. | Incorporates a robust measurement system for assessing results and organizational performance. | Recommends the use of performance indicators and monitoring to drive improvement. |
| Benchmarking and Best Practices | Encourages benchmarking and learning from best practices within and outside the organization. | Incorporates benchmarking as a means to achieve excellence and innovation. | Promotes benchmarking and the adoption of best practices for performance improvement. |
| Stakeholder Focus | Emphasizes understanding and meeting the needs of stakeholders. | Central to the model, considering the interests and needs of all stakeholders. | Acknowledges the importance of understanding and meeting the needs of relevant stakeholders. |
| Integration with ISO 9001 | Complementary to ISO 9001, providing guidance for sustained success beyond basic compliance. | Can be integrated with ISO 9001 but extends beyond its requirements. | Provides guidance for organizations already implementing ISO 9001. |

*Source:* Own elaboration based on: (ISO 9004:2018; ISO ISO 10014:2021).

## Maturity Model according to ISO 9004

ISO 9004 is a standard developed by the International Organization for Standardization (ISO) that provides guidelines for managing the sustained success of an organization. The MM described in ISO 9004 is designed to help organizations assess and improve their overall performance by focusing on the maturity of their management system. ISO 9004 standard also allows for self-assessment of the maturity level of the quality management system, as well as the improvement of the entire organization on a five-level scale (Glogovac et al., 2022).

The MM in ISO 9004 is based on a set of maturity levels, each representing a stage of development in the organization's management system. These levels are defined in terms of key characteristics and practices that contribute to sustained success. The model typically consists of five maturity levels, often described as follows (ISO 900:2018):

- **Initial Level:** At this stage, the organization has a basic understanding of its processes and their interactions. There may be ad-hoc efforts to improve performance, but there is a lack of systematic management.
- **Managed Level:** The organization starts to establish basic processes and controls to manage its key activities. Documentation and standardization of processes may begin, and there is a growing awareness of the need for improvement.
- **Established Level:** At this point, the organization has well-defined and documented processes in place. It demonstrates a commitment to continuous improvement, with a focus on monitoring and measuring performance. The organization may also have established goals and objectives for improvement.
- **Predictable Level:** The organization's processes are now well-established and consistently applied. There is a strong emphasis on data-driven decision-making, and the organization can predict its performance with a high degree of confidence. Continuous improvement is ingrained in the organizational culture.
- **Optimizing Level:** This is the highest level of maturity, where the organization continually seeks opportunities for innovation and optimization. The focus is on achieving breakthrough performance and anticipating future challenges. The organization actively engages in benchmarking, best practices, and proactive risk management.

Table 11 describes of main features of the five levels of maturity.

**Table 11**  Levels of maturity of quality management system functioning according ISO 9004:2018.

| Level | Description |
|---|---|
| **1. Initial** | • A very basic level of quality management system awareness and implementation,<br>• Processes are informal and ad-hoc,<br>• There is no systematic approach to quality management,<br>• Quality is not a strategic priority. |
| **2. Repeatable** | • Quality management system processes are documented and standardized,<br>• Processes are followed consistently, but there may be some variations,<br>• Quality management is considered important, but there is no formal approach to improvement. |
| **3. Defined** | • Quality management system processes are well-defined and integrated into the organization's operations,<br>• There is a clear understanding of the organization's quality policy and objectives,<br>• Quality management is embedded in the organization's culture. |

*Contd.*

**Table 11** *Contd.*

| Level | Description |
|---|---|
| **4. Managed** | • Quality is proactively managed and improved<br>• Data is used to identify and address quality issues<br>• Risks are managed to prevent problems from occurring |
| **5. Optimizing** | • Quality management is continuously improved based on data-driven insights,<br>• The organization is constantly innovating and seeking new opportunities to improve its quality management system,<br>• Quality is a key driver of the organization's success. |

*Source:* Own elaboration based on: (ISO 9004:2018).

Despite somewhat different origins related to quality management systems, it is also successfully applied in practice to assess the maturity level of business processes and improvement in the selected area. Quality management is fundamentally based on a process approach. ISO 9004 standard even recommends that the strategic goal of the organization should be continuous improvement of processes, conducted to improve the functioning of the organization and increase benefits for all interested parties (Mazur, 2014).

The MM outlined in ISO 9004 is intended to be flexible, allowing organizations to adapt it to their specific context and industry. The goal is to help organizations move through the maturity levels systematically, with the ultimate aim of achieving sustained success and long-term viability. The model also encourages a holistic approach, integrating quality management with overall business strategy and objectives.

Implementing an MM according to ISO 9004:2018 is a strategic process that requires commitment from the senior management and a team of dedicated individuals. It is important to have a clear understanding of the organization's current level of maturity, and to set realistic goals for improvement (Moradi-Moghadam et al., 2013).

The steps in implementing an MM according to ISO 9004:2018 are:

- **Assess the organization's current level of maturity:** There are a number of tools and methods that can be used to assess an organization's current level of maturity. One approach is to use the ISO 9004:2018 MM itself. This model can be used to evaluate an organization's performance against the five levels of maturity.
- **Set clear goals for improvement:** Once the organization's current level of maturity has been assessed, the next step is to set clear goals for improvement. These goals should be SMART: Specific, Measurable, Achievable, Relevant, and Time-bound.
- **Develop a plan to achieve the goals:** Once the goals have been set, a plan needs to be developed to achieve them. This plan should identify the specific actions that need to be taken, who will be responsible for taking them, and when they will be completed.

- **Implement the plan:** The plan should then be implemented. This will involve taking the actions that have been identified and tracking progress towards the goals.
- **Monitor and evaluate progress:** It is important to monitor and evaluate progress on a regular basis. This will help to ensure that the organization is on track to achieve its goals.
- **Make adjustments as needed:** If the organization is not making progress, or if it is not on track to achieve its goals, it may be necessary to make adjustments to the plan.

Table 12 has a description of main benefits of implementation of an MM according to ISO 9004:2018 in organizations.

**Table 12**    Benefits of implementation of maturity model according to ISO 9004:2018.

| Benefit | Description |
| --- | --- |
| Improved quality | A maturity model can help an organization to improve its quality by providing a framework for identifying and addressing quality problems. |
| Increased efficiency | A maturity model can help an organization to increase its efficiency by streamlining processes and eliminating waste. |
| Reduced costs | A maturity model can help an organization to reduce its costs by reducing the number of defects and rework. |
| Enhanced customer satisfaction | A maturity model can help an organization to enhance customer satisfaction by providing a better quality product or service. |

*Source:* Own elaboration based on: (ISO 9004:2018).

Implementing an MM according to ISO 9004:2018 is a valuable tool for organizations that want to improve their quality management systems and achieve sustained success. The model (Table 13) can help organizations to identify areas for improvement, develop a plan to achieve their goals, and monitor and evaluate their progress. By following the steps outlined in this article, organizations can successfully implement an MM and reap the rewards of improved quality, increased efficiency, and reduced costs.

**Table 13**    Main features of maturity model according to ISO 9004:2018.

| Feature | Description |
| --- | --- |
| Commitment from senior management | A strong commitment from senior management is essential for successful implementation of a maturity model. Without this commitment, it will be difficult to gain the resources and support that are necessary to make the model work. |
| A team of dedicated individuals | A team of dedicated individuals is needed to implement the maturity model. This team should include representatives from all levels of the organization and should have the skills and experience necessary to lead the implementation effort. |

*Contd.*

**Table 13**  *Contd.*

| A clear understanding of the organization's current level of maturity | It is important to have a clear understanding of the organization's current level of maturity before implementing a maturity model. This can be done by using a maturity model assessment tool or by conducting a self-assessment. |
|---|---|
| Realistic goals for improvement | The goals for improvement should be SMART: Specific, Measurable, Achievable, Relevant, and Time-bound. This will help to ensure that the organization is able to achieve its goals. |
| A well-defined plan | A well-defined plan is essential for successful implementation of a maturity model. The plan should identify the specific actions that need to be taken, who will be responsible for taking them, and when they will be completed. |
| Regular monitoring and evaluation | The organization should monitor and evaluate its progress on a regular basis. This will help to ensure that the organization is on track to achieve its goals. |
| Flexibility | The maturity model should be flexible enough to accommodate the needs of the organization. The organization may need to modify the model to fit its specific circumstances. |
| Continuous improvement | The organization should continuously improve its maturity model. This will help the organization to maintain its competitive edge. |

*Source:* Own elaboration based on: (ISO 9004:2018).

## Maturity Model according to ISO 10014:2021

This document provides guidance on achieving financial and economic benefits through the implementation of a systematic top-down approach. This structured approach aligns with the quality management principles and system outlined in the ISO 9000 family of management system standards. The key objectives are:

- Monitoring and managing trends in crucial performance metrics.
- Taking improvement actions based on the observed metrics.

The maturity assessment model aims to measure the functioning of an organization's quality management system in terms of achieving economic and financial benefits (Moradi et al., 2015). One of the methods used in this area is measurement according to the model described in the PN-ISO 10014:2022 standard Quality management. Guidelines for achieving financial and economic benefits. This model is a universal approach as it can be applied to any type of organization. The standard is applicable to organizations whose products include services, intellectual products, material objects, and processed materials. It can be applied in both the public and private sectors and can provide useful guidance, regardless of the number of employees, the diversity of products offered, income, process complexity, or the number of locations.

The intended audience for this document is specifically the top management of organizations. It is applicable to organizations of any type, including public, private, or not-for-profit sectors, irrespective of their business model, revenue, employee count, product and service diversity, organizational culture, process complexity, location, or number of locations (Wolniak, 2013).

The ISO 10014:2021 standard serves as a supplement to ISO 9001:2015 and ISO 9004:2018, focusing on performance improvements. It provides practical examples of achievable benefits derived from applying the concepts in those standards. Additionally, the document outlines practical management methods and tools to support organizations in realizing these benefits.

The ISO 10014:2021 standard defines an MM that can be used to assess an organization's ability to effectively implement and maintain a quality management system. The model consists of five levels, each representing a different level of maturity. Their description is in Table 14. Main features of this model are outlined in Table 15.

**Table 14**　Levels of maturity according to ISO 10014:2021.

| Level | Description |
|---|---|
| **Level 1. Unmanaged** | Organizations at this level have a limited understanding of quality management or may not even be aware of the concept. They may have no formal quality management system in place, or if they do, it may be poorly documented and implemented. |
| **Level 2. Reactive** | Organizations at this level recognize the importance of quality and may have a basic quality management system in place. However, their approach is reactive rather than proactive. They may only focus on quality when there is a problem, and they may not have a systematic approach to improvement. |
| **Level 3. Proactive** | Organizations at this level have a more proactive approach to quality management. They have a documented quality management system that is implemented and maintained consistently. They also use data and analysis to identify and address quality issues before they become problems. |
| **Level 4. Integrated** | Organizations at this level have integrated quality management into their overall business strategy. Quality is a core value of the organization, and it is embedded in all aspects of their operations. They use quality management to improve customer satisfaction, reduce costs, and achieve their business goals. |
| **Level 5. Excellence** | Organizations at this level are leaders in quality management. They are constantly innovating and seeking new ways to improve their quality management system. They are also recognized for their quality by their customers, suppliers, and other stakeholders. |

*Source:* Own elaboration based on: (ISO 10014:2021).

**Table 15**   Main features of maturity model according to ISO 10014:2021.

| *Feature* | *Description* |
|---|---|
| Focus on quality management as a strategic enabler | The ISO 10014:2021 maturity model emphasizes the importance of quality management as a strategic enabler for organizations to achieve their overall business objectives. |
| Alignment with organizational strategy | The model encourages organizations to align their quality management system with their overall business strategy to ensure that quality initiatives are focused on achieving strategic goals. |
| Integration of quality management into organizational processes | The model promotes the integration of quality management into all aspects of an organization's operations, from product design and development to customer service and supplier management. |
| Data-driven decision-making | The model emphasizes the importance of using data and analysis to drive continuous improvement in the quality management system. |
| Continuous improvement culture | The model encourages organizations to cultivate a culture of continuous improvement where employees are empowered to identify and address quality issues proactively. |
| Organizational learning | The model emphasizes the importance of organizational learning as a key enabler of sustained quality improvement. |

*Source:* Own elaboration based on: (ISO 10014:2021).

**Technological Maturity Scale**

The concept of maturity can be used in analysis of many problems. For example, a very interesting concept is called Technological Maturity Scale or Technological Maturity Level (TML).

The Technology Readiness Level (TRL) serves as a metric to characterize a technology's developmental stage and maturity. Originating in the 1970s at NASA, it initially aimed to gauge a technology's proximity to deployment in space, helping to establish minimum flight readiness requirements for components to mitigate risks in future missions (Boretti, 2023).

The U.S. Department of Defense later integrated TRL into the procurement of weapons technology, leading to its adoption by various governmental, military organizations, and the European Space Agency. The European Union introduced the concept of LRT in 2009, emphasizing Key Enabling Technologies (KETs) in a communication titled "Preparing our Future". Unlike the traditional application of TRL in gauging a technology's readiness for operational use, the EU's interpretation extends to the readiness of a product or service for commercialization (Salazar, 2022).

This broader interpretation arose from a study by the High-Level Expert Group on Key Enabling Technologies, identifying the "Valley of Death" as a significant obstacle to technology transfer within the EU. The Valley of Death represents a bottleneck in technological innovation projects, occurring after research and development phases but before the launch of new products to the market. The EU's predominant focus on low TRLs, particularly in basic and applied science projects, poses challenges in scaling up results to market-level processes, technologies, or services (TRL 9).

The difficulty lies in the translation of basic research into viable commercial applications. Therefore, employing the TRL as a reference helps identify the level of advancement and maturity of technologies, determining project feasibility and the proximity of a product to the market (Santos et al., 2023). The levels of technology readiness are explained in Table 16.

**Table 16**  Technology Readiness Levels.

| Level | Description |
| --- | --- |
| TRL 1 | At this stage, technology is in its infancy, with basic principles just observed and reported. There is no practical application identified, and the focus is on fundamental scientific research. |
| TRL 2 | The technology concept begins to take shape at this level, with potential applications identified. However, it is still in the early stages, and practical feasibility is uncertain. Research efforts are directed towards refining the concept and understanding its potential. |
| TRL 3 | Moving beyond theoretical concepts, TRL 3 involves demonstrating critical functions and providing proof-of-concept through analytical and experimental means. The emphasis is on validating the fundamental ideas and gaining confidence in their viability. |
| TRL 4 | Validation efforts progress to the laboratory environment, where components and subsystems are tested to ensure they meet the intended specifications. This stage is crucial for identifying technical challenges and refining the design before moving forward. |
| TRL 5 | Building on laboratory validation, TRL 5 involves testing components and subsystems in a relevant environment. This step aims to assess the technology's performance under conditions that mimic real-world scenarios, providing valuable insights for further development. |
| TRL 6 | Advancing to the prototype stage, TRL 6 sees the development and demonstration of a system prototype. This prototype is tested in a relevant environment to evaluate its overall functionality and integration, addressing any unforeseen issues that may arise. |
| TRL 7 | Technology reaches a more mature stage at TRL 7, where the system prototype is demonstrated in an operational environment. This stage involves testing the technology in conditions closely resembling its intended operational setting, ensuring its readiness for real-world use. |
| TRL 8 | At this level, the technology undergoes rigorous testing to qualify its components and systems. The focus is on validating the actual system, ensuring it meets all requirements and standards. TRL 8 marks the transition from development to production readiness. |
| TRL 9 | The highest level on the scale, TRL 9 signifies the technology's proven success through actual mission operations. It has been deployed, operated, and validated in real-world scenarios, demonstrating its reliability, effectiveness, and readiness for widespread use. |

*Source:* Own elaboration based on: (Salazar, 2022; Santos et al., 2023; Englezos, 2023; Seipolt et al., 2023).

A methodical approach to Technology Readiness Levels (TRLs) is necessary to guide a technology's progression from conception, through research and development to deployment. Universities and government funding primarily emphasize TRLs 1–4, while the private sector directs its attention to TRLs 7–9. In Table 17, the main advantages of technology readiness' level usage are elucidated.

**Table 17**  The advantages of usage of Technology Readiness Levels.

| *Advantage* | *Description* |
|---|---|
| Objective Assessment | TRL provides a standardized and objective way to assess the maturity of a technology, reducing subjective biases. |
| Risk Mitigation | Helps in identifying and mitigating risks associated with the development and implementation of new technologies. |
| Project Planning | Enables better project planning by offering a clear understanding of the technology's current state and future needs. |
| Resource Allocation | Facilitates informed decision-making regarding resource allocation based on the technology's readiness level. |
| Communication and Collaboration | Enhances communication between stakeholders by providing a common language to discuss the maturity of technologies. |
| Funding Prioritization | Assists funding agencies in prioritizing projects based on their technological readiness and potential impact. |
| Technology Transfer | Simplifies the transfer of technology from research and development to practical applications in various industries. |
| Quality Control | Supports quality control processes by ensuring that technologies are sufficiently mature before widespread adoption. |
| Informed Investment Decisions | Facilitates well-informed investment decisions by investors and businesses, reducing uncertainties and improving ROI. |
| Regulatory Compliance | Aids in complying with regulatory requirements by demonstrating the maturity and safety of technologies under consideration. |
| Facilitates Collaboration with Industry | Encourages collaboration between researchers, industry partners, and policymakers by providing a common assessment framework. |

*Source:* Own elaboration based on: (Salazar, 2022; Santos et al., 2023; Englezos, 2023; Ahmad, Sabtacruz et al., 2023; Lavin et al., 2022).

## 3.  Holistic Approach in Industrial Transformation

The holistic approach in industrial transformation signifies a paradigm shift from isolated, siloed strategies to a comprehensive and integrated methodology that encompasses all facets of an organization's operations. In the context of industrial transformation, this approach goes beyond the mere adoption of new technologies or process improvements; it extends to the entire ecosystem of the industrial landscape. A holistic perspective recognizes that industrial transformation is not solely about implementing advanced technologies or automating processes. It emphasizes the need for a synchronized and interconnected strategy that considers the integration of technology, human capital, organizational culture, and environmental sustainability. By doing so, organizations can navigate the complexities of modern industrial environments more effectively.

In the area of technology, a holistic approach involves the strategic adoption of innovations like IoT, AI, and advanced analytics. It does not view technology in isolation. Instead, it considers how these technological advancements align with the broader goals of the organization, enhance collaboration, and contribute to a more resilient and adaptive operational framework.

Human capital is another integral component of the holistic industrial transformation approach. Recognizing that successful transformation requires a skilled and motivated workforce, organizations focus on upskilling employees, fostering a culture of innovation, and creating environments that encourage collaboration between humans and machines. This people-centric perspective ensures that technological advancements are harnessed to augment human capabilities rather than replace them.

Cultural and organizational aspects are also important in a holistic approach. Organizations must cultivate a culture that values continuous learning, embraces change, and encourages cross-functional collaboration. The alignment of organizational goals with the transformation strategy becomes crucial, ensuring that everyone within the organization is working towards a common vision.

It can be stated that, a holistic perspective extends to environmental sustainability. Organizations are increasingly recognizing the importance of incorporating eco-friendly practices into their industrial transformation efforts. This includes optimizing energy consumption, minimizing waste, and adopting environmentally responsible manufacturing processes. The aim is not only to enhance operational efficiency but also to contribute positively to the broader environmental landscape.

Holistic MMs provide organizations with a comprehensive framework for assessing their overall development across various dimensions. Unlike traditional MMs that focus on specific aspects such as technology or processes, holistic MMs take a more inclusive approach, considering the entire organizational ecosystem. Table 18 shows a comparison of traditional and holistic MMs.

**Table 18** Comparison between holistic and traditional maturity models.

| Aspect | Holistic Maturity Models | Traditional Maturity Models |
|---|---|---|
| Scope | Holistic maturity models take a broad and comprehensive view, considering the entire organizational ecosystem. This includes people, processes, technology, culture, and external factors. The scope extends beyond isolated components to capture the interconnectedness of various elements within the organization. | Traditional maturity models typically focus on specific areas such as technology, processes, or certain organizational functions. They have a more limited scope and may not consider the broader organizational ecosystem. |

*Contd.*

**Table 18**  *Contd.*

| *Aspect* | *Holistic Maturity Models* | *Traditional Maturity Models* |
|---|---|---|
| Integration | Holistic models emphasize the integration of different dimensions, recognizing that advancements in one area may impact others. This integrated approach ensures a synchronized development across various aspects of the organization. | Traditional models often treat components in isolation, lacking a holistic integration perspective. They may assess and improve specific elements without considering their interdependencies with other organizational facets. |
| Viewpoint | Holistic maturity models provide a panoramic view of an organization's evolution. They consider the dynamic interactions between different elements and offer a nuanced understanding of its overall health and development. | Traditional maturity models provide a more focused and narrow view, concentrating on specific elements without necessarily considering their broader impact on the entire organization. |
| Assessment | Holistic models consider both internal and external factors influencing organizational maturity. This includes aspects like strategic alignment, leadership effectiveness, employee engagement, and adaptability to change. | Traditional models primarily assess internal factors related to a specific area, such as technology or processes. External factors may not be explicitly considered in the assessment process. |
| Adaptability | Holistic models advocate for an ongoing journey of continuous improvement and adaptation to change. They recognize that organizational maturity is dynamic and requires a proactive approach to navigate evolving market conditions. | Traditional models may be more static, with periodic assessments and updates. The focus might be on achieving a particular maturity level rather than fostering continuous adaptability. |
| Mindset | Holistic models encourage an adaptive mindset, viewing maturity as a continuous journey rather than a final destination. This mindset promotes a culture of resilience and openness to ongoing improvement. | Traditional models may imply a goal-oriented mindset with a defined endpoint. Once a specific maturity level is reached, the emphasis may shift to maintaining that level rather than continuous progression. |
| Culture | Holistic models highlight the importance of cultivating a culture that fosters innovation, collaboration, and continuous improvement. This cultural aspect is considered integral to achieving and sustaining maturity. | Traditional models may not explicitly address cultural aspects in their assessments. The focus might be more on procedural adherence and efficiency rather than fostering a specific organizational culture. |
| Agility | Holistic models promote resilience and agility as essential attributes in the face of uncertainty and change. They emphasize the need for organizations to adapt quickly and effectively to evolving circumstances. | Traditional models may not explicitly focus on building agility into the model. The emphasis might be on achieving a predetermined level of maturity without a specific focus on organizational agility. |

*Contd.*

**Table 18** *Contd.*

| *Aspect* | *Holistic Maturity Models* | *Traditional Maturity Models* |
|---|---|---|
| Human Element | Holistic models recognize the importance of the human factor in transformation. They emphasize upskilling employees, fostering a culture of innovation, and creating environments that encourage collaboration between humans and technology. | Traditional models may focus more on processes and technology, potentially neglecting the human aspects of transformation. The role of the workforce in driving and sustaining maturity may not be as explicitly highlighted. |
| Environmental Sustainability | Holistic models incorporate eco-friendly practices into industrial transformation efforts. This includes optimizing energy consumption, minimizing waste, and adopting environmentally responsible manufacturing processes. | Traditional models may not explicitly consider environmental factors in their assessments. The focus may be primarily on internal efficiency and effectiveness without a specific emphasis on sustainability practices. |

*Source:* Own elaboration.

These models recognize that true maturity extends beyond individual components and necessitates a synergistic integration of people, processes, technology, and culture. They aim to capture the interconnectedness of these elements, acknowledging that advancements in one area may impact others. In essence, holistic MMs offer a panoramic view of an organization's evolution, enabling a nuanced understanding of its strengths and areas that require improvement.

Organizational maturity is often assessed through multiple lenses, such as strategic alignment, leadership effectiveness, employee engagement, and adaptability to change. Holistic MMs go beyond surface-level evaluations, delving into the underlying dynamics that contribute to an organization's overall health. They emphasize the importance of cultivating a culture that fosters innovation, collaboration, and continuous improvement.

Furthermore, these models recognize that achieving maturity is an ongoing journey rather than a destination. They advocate for iterative assessments and adjustments, encouraging organizations to adapt to evolving market conditions, technological advancements, and societal changes. This adaptive mindset is integral to holistic maturity, as it promotes resilience and agility in the face of uncertainty.

Holistic MMs also highlight the interconnected nature of internal and external factors influencing organizational growth. Factors such as customer satisfaction, market trends, and regulatory compliance are considered alongside internal metrics, creating a more nuanced understanding of an organization's position in its ecosystem.

Industry 4.0 maturity is conceptualized as a holistic model that evaluates an organization's preparedness and advancement in embracing Industry 4.0 technologies. This comprehensive model considers various dimensions to gauge the organization's transition towards a smart and interconnected ecosystem (Bajic et al., 2023).

The Industry 4.0 MM serves as a measurable gauge for evaluating the effects of the transformation in business operations and maturity. Industry 4.0 entails a shift from traditional to digital operations, aiming for heightened visibility, transparency, and efficiency (Mittal et al., 2018; Onueme and Liyanage, 2023). The advent of digital technologies such as IoT, AI, Edge, and Cloud Computing form the cornerstone of this industrial transformation. However, it's essential to recognize that transformation extends beyond technology; it entails a harmonious integration of people adoption, process alignment, and technological advancement (Ganesh, 2023).

Achieving the full benefits of Industry 4.0 necessitates an incremental approach to transformation, progressively advancing through the stages of the Industry 4.0 MM—The Manufacturing Enterprise 4.0 Maturity Model. Those stages are described in Table 19.

One critical dimension is technological infrastructure, encompassing aspects such as connectivity, integration, and sensorization. The organization's ability to manage and derive insights from data is another crucial dimension, incorporating elements like data collection, analytics, and predictive maintenance. The cyber-physical systems dimension evaluates the level of automation, the integration of robotics, and the implementation of digital twin technology. Human-machine interaction is also examined, including the use of augmented reality, virtual reality, and collaborative robots (Treviño-Elizondo and Garcia-Reyes, 2023; Dal Forno et al., 2023).

Supply chain integration is a significant aspect, focusing on end-to-end visibility and collaboration with partners. Organizational culture and skill development are evaluated, considering the establishment of a digital culture and the enhancement of employee skills related to Industry 4.0 technologies (Semerano et al., 2023; Henriquez et al., 2023). Security and risk management are vital dimensions, encompassing cybersecurity measures and risk assessment capabilities. Business model innovation is assessed, exploring new models like product-as-a-service and servitization. Regulatory compliance, adherence to standards, and alignment with industry regulations constitute another crucial dimension.

This holistic approach ensures a thorough evaluation of the organization's Industry 4.0 maturity, helping identify strengths, weaknesses, and areas for improvement in its journey towards a technologically advanced future.

Manufacturing plants have the opportunity to significantly enhance their productivity through the adoption of digital manufacturing. However, not all facilities are equally prepared to leverage this technology. Those that can advance to more advanced stages of digital maturity currently have the opportunity to outperform their counterparts and secure a larger portion of their industry's customer base.

**Table 19**  Stages of the Manufacturing Enterprise 4.0 Maturity Model.

| *Stage* | *Description* |
| --- | --- |
| Start with Asset Connectivity | Initiating the journey into Industry 4.0 involves closing the divide between the physical and digital realms. In the industrial context, this entails the integration of Operational Technology (OT) and Information Technology (IT). The connectivity facilitated by the Internet of Things (IoT) plays a pivotal role in linking machinery, sensors, Programmable Logic Controllers (PLCs), and SCADA systems to enterprise systems, applications, and infrastructure through an IoT Platform. The deployment of IoT Gateways and Edge Devices is instrumental in filtering out unnecessary data from machines, ensuring that only refined data is processed. |
| Consolidate data to access from anywhere / anytime | The conventional method of aggregating and archiving data at the factory floor (on-premise) constrains the accessibility, visibility, and insights derived within the premises. Retrieving this data for decision-makers involves manual processes, causing delays and hindering effective planning and timely business optimization. Leveraging the capabilities of cloud computing empowers business users to amalgamate machine data, business systems, and various events, consolidating them in the cloud for swift processing and enhanced visibility. Classic technologies like Data-warehouse, DataLake, or Lighthouse are available for the consolidation of industrial data. |
| Perform analysis to understand "what happened" | Once all the data collected from the factory is accessible in the cloud, the next step involves analyzing the data to derive more profound insights and intelligence to assist business users. The essence of Industry 4.0 lies in data-driven business insights and intelligence that aid decision-makers in aligning and optimizing business operations to enhance profitability. Manufacturing intelligence solutions provide crucial performance indicators like OEE, TEEP, Time Loss, and Takt value, motivating factory leaders to enhance floor operations. |
| Predict to understand "what will happen in the future" | Envisioning the future has perennially been an aspiration for individuals, particularly business leaders. Having a technology that enables the prediction of "What will happen in the near future?" serves as an invaluable tool for future planning. Modern tools such as Artificial Intelligence and Machine Learning (ML) empower industry leaders to anticipate forthcoming challenges and issues based on accumulated data. The fourth stage of Industry 4.0 maturity involves harnessing data to train ML models, facilitating predictions related to machine maintenance, quality concerns, demand trend analysis, supply chain obstacles, and operational downtime. |

*Contd.*

**Table 19** *Contd.*

| Stage | Description |
|---|---|
| Train machines to act autonomously "when it happens" | Industrial automation originated in the era of Industry 3.0. While robotic automation in this phase contributed to minimizing human efforts and optimizing production in high volumes, it remained limited in accessibility and control within the factory setting. In the present digital age and interconnected world, there is a need to upgrade factories for autonomous operation to align with global market demands. By incorporating IoT, AI, and ML technologies to foresee future factory scenarios, these technologies can also be employed to establish autonomous operations seamlessly integrated with existing automation systems. This advancement paves the way for Industry 5.0, representing the ultimate level of maturity in digital transformation for the industry. |

*Source:* Own elaboration based on: (Ganesh, 2023).

Table 20 presents the Industry 4.0 maturity levels.

**Table 20**　Industry 4.0 maturity levels.

| Stage | Description |
|---|---|
| **Level 1. Computerization** | At this stage, companies are using digital solutions like CNC machining. However, these solutions are all operating in isolation. They are not connected to each other, and there is no way of coordinating a combined effort between them without significant manual input. Any production data they collect is also individualized, siloed, and only available after production has already concluded. |
| **Level 2. Connectivity** | Companies progress to this phase when their digital solutions exhibit some level of interconnectivity. Manufacturers had the opportunity to reach this stage as early as the beginning of the 2000s by linking their machines to the internet. While this stage permits some communication between machines, it is typically characterized by sluggish and inefficient processes. The connections lack the agility to keep pace with real-time production events, resulting in a disparity between the data and the actual occurrences on the factory floor. |
| **Level 3. Visibility** | By the third stage of this process, manufacturers have overcome this bottleneck and now have access to a "digital twin" of their factory made with comprehensive real-time data. This data is detailed enough to let observers know exactly what is going on at the factory at any given time, but it may lack the organization necessary to make this possible. At this point, the presence of the data itself is the only qualifying factor. |

*Contd.*

**Table 20** *Contd.*

| *Stage* | *Description* |
|---|---|
| **Level 4. Transparency** | During this phase, the focus of digital advancement transitions from enhancing technical capabilities to harnessing analytics. To reach this stage, manufacturers need to acquire the skill of interpreting and comprehending the data they have gathered. Many shop floor control software systems carry out this type of analysis automatically, organizing and presenting pertinent data in response to user queries. |
| **Level 5. Predictive Capacity** | In this stage, manufacturers utilize the available real-time data to forecast future developments in their production capabilities. This involves anticipating events like potential work stoppages, identifying potential quality control issues, pinpointing opportunities for energy conservation within the facility, and more. Shop floor control software plays a substantial role in facilitating these predictions. |
| **Level 6. Adaptability** | In this ultimate phase, the digital systems within the facility attain the capacity to autonomously operate based on their analytical and predictive capabilities. They can proactively respond to evolving conditions by coordinating staff and their responsibilities, effectively managing both routine operations and addressing unexpected emergencies and emerging issues. |

*Source:* Own elaboration based on: (The Industry, 2021; Istaitih, 2020).

Another MM of Industry 4.0 is SIMMI 4.0. The SIMMI 4.0 Maturity Model represents a comprehensive framework designed to assess and enhance an organization's maturity in the context of Industry 4.0. This model, standing for Smart, Integrated, Mature, Modular, and Innovative, serves as a guide for businesses navigating the complexities of the Fourth Industrial Revolution (Leyh and Schäffer, 2018).

In the Smart dimension, organizations focus on leveraging advanced technologies like AI, ML, and the IoT to make data-driven decisions. This involves the integration of smart technologies into various aspects of operations, leading to increased efficiency and responsiveness (Leyh et al., 2017).

Integration is a key element of the SIMMI 4.0 MM. It emphasizes the need for seamless connectivity across different processes, departments, and systems. Integrated systems enable real-time information sharing, facilitating a more holistic and synchronized approach to organizational functions (Souhail et al., 2023).

Maturity in the context of SIMMI 4.0 involves the continuous improvement and optimization of processes and technologies. Organizations strive to reach higher levels of maturity by refining their strategies, adopting best practices, and staying abreast of emerging trends in the rapidly evolving landscape of Industry 4.0. A description of dimensions of SIMMI 4.0 model is presented in Table 21.

**Table 21**  Dimensions of SIMMI 4.0 Maturity Model.

| Dimension | Description |
|---|---|
| Smart | The Smart dimension focuses on leveraging advanced technologies such as artificial intelligence, machine learning, and the Internet of Things. Organizations strive to make data-driven decisions and integrate smart technologies into various aspects of their operations to enhance efficiency and responsiveness. |
| Integrated | Integrated emphasizes the need for seamless connectivity across different processes, departments, and systems within an organization. This dimension encourages real-time information sharing, creating a more holistic and synchronized approach to organizational functions. |
| Mature | The Mature dimension revolves around the continuous improvement and optimization of processes and technologies. Organizations aim to reach higher levels of maturity by refining strategies, adopting best practices, and staying updated on emerging trends in the rapidly evolving landscape of Industry 4.0. |
| Modular | Modular encourages organizations to adopt a flexible and scalable approach to system design. A modular approach allows for easy integration of new technologies and adaptation of existing ones, ensuring that systems can evolve to meet changing business requirements efficiently. |
| Innovative | Innovative underscores the importance of fostering a culture of innovation within the organization. This dimension involves encouraging creativity, experimentation, and a proactive approach to embracing emerging technologies. Organizations at the highest level of maturity continuously seek and implement cutting-edge solutions. |

*Source:* Own elaboration based on: (Leyh and Schäffer, 2018; Leyh et al., 2017; Leyh et al., 2016).

Modularity is another critical aspect, encouraging organizations to build flexible and scalable systems. A modular approach allows for the easy integration of new technologies and the adaptation of existing ones, ensuring adaptability in the face of changing business requirements (Martins et al., 2023).

The Innovative dimension underscores the importance of fostering a culture of innovation within the organization. This involves encouraging creativity, experimentation, and a proactive approach to embracing emerging technologies. Organizations at the highest level of maturity in this dimension continuously seek out and implement cutting-edge solutions to stay ahead in the competitive landscape (Leyh et al., 2016).

Another interesting Industry 4.0 MM is The Manufacturing Enterprise 4.0 Maturity Model (MEMM). It is a framework that helps organizations assess their level of readiness for Industry 4.0 (Simetinger and Basi, 2022). It was developed by the German Federal Ministry of Education and Research (BMBF) and is based on the principles of the Capability Maturity Model Integration (CMMI).

The MEMM consists of five maturity levels, each of which represents a different level of maturity in terms of Industry 4.0 capabilities. The levels are presented in the Table 22.

**Table 22** Levels of The Manufacturing Enterprise 4.0 Maturity Model (MEMM).

| *Dimension* | *Description* |
|---|---|
| **Adoption** | Organizations at this level are in the early stages of understanding Industry 4.0 and its potential impact. They may have some awareness of key concepts and technologies, but their understanding is still limited. They have not yet taken any concrete steps to implement Industry 4.0 technologies. |
| **Familiarity** | Organizations at this level have a basic understanding of Industry 4.0 concepts and have initiated initial steps towards adoption. They may have conducted some exploratory research, attended industry events, or engaged with consultants. They have implemented a few isolated Industry 4.0 projects, such as pilot deployments of predictive maintenance or machine learning applications for quality control - Conceptual understanding of Industry 4.0 concepts. |
| **Deployment** | Organizations at this level are actively using Industry 4.0 technologies to improve their operations across multiple departments and processes. They have established a foundation of Industry 4.0 infrastructure and are leveraging technologies such as sensors, data analytics, and automation to enhance efficiency, productivity, and quality. However, Industry 4.0 technologies are still not fully integrated into their core business processes. |
| **Integration** | Organizations at this level have seamlessly integrated Industry 4.0 technologies into their core business processes. They have achieved a holistic view of their operations, enabling them to optimize their entire value chain using interconnected Industry 4.0 solutions. Industry 4.0 technologies are embedded in the fabric of the organization and are used to make data-driven decisions at all levels. |
| **Transformation** | Organizations at this level have undergone a fundamental transformation of their business models and operations to fully embrace Industry 4.0 principles. They have developed a new business model that is driven by data and digital technologies, enabling them to adapt to changing market conditions and customer demands. Their organizational structure and culture have also been transformed to support the agility and innovation required for Industry 4.0 success - Industry 4.0 is not just a set of technologies; it is a way of doing business. |

*Source:* Own elaboration based on: (Simetinger and Basi, 2022; Schumacher et al., 2019).

A comparison of main important Industry 4.0 MMs is elucidated in Table 23. Using the criteria such as, origin of model, it's focus, maturity levels, and key dimension, we tried to juxtapose them to observe the similarities and differences of them.

Holistic models used to measure the level of maturity are also digital maturity models (Skalli et al., 2023). A digitally mature company is an establishment that, in all its processes and products, is aware of the solutions available in the market and utilizes those that bring the greatest value to its business. In other words, it

**Table 23** The characteristic of holistic Industry 4.0 maturity models.

| Maturity Model | Origin | Focus | Maturity Levels | Key Dimensions |
|---|---|---|---|---|
| **Manufacturing Enterprise 4.0 Maturity Model (MEMM)** | German Federal Ministry of Education and Research (BMBF) | Overall readiness for Industry 4.0 adoption | Adoption, Familiarity, Deployment, Integration, Transformation | Digital Infrastructure, Data Management, Analytics, Process Automation, Human Capital |
| **System Integration Maturity Model Industry 4.0 (SIMMI 4.0)** | TU Dresden | Framework for assessing an organization's IT system landscape in terms of Industry 4.0 requirements | Foundation, Standardization, Integration, Optimization, Transformation | System Integration, Data Management, Analytics, Interoperability, and Leadership |
| **Industry 4.0 Maturity Index (I4.0MI)** | Matics | Assessment of Industry 4.0 capabilities across various areas | Basic, Intermediate, Advanced, Expert | Strategic Alignment, Digital Transformation, Data and Analytics, Automation, Ecosystem |

*Source:* Own elaboration based on: (Leyh and Schäffer, 2018; Leyh et al., 2017; Leyh et al., 2016; Simetinger and Basi, 2022; Schumacher et al., 2019; Ganesh, 2023; The Industry, 2021; Istaitih, 2020).

**Table 24** Digital Maturity Model in Industry 4.0.

| Module | Pillars | Example |
|---|---|---|
| **Technologies** | Automation | Automation is to be understood as the replacement of repetitive tasks performed by humans with the work of programmed machines capable of reacting to stimuli during the execution of processes. In an automated environment, human work is reduced to supervizing machine operations, possibly responding to certain process phenomena and issues. Automation is achieved through the implementation of technologies for monitoring, controlling, and executing production, as well as delivering goods and services. It not only frees employees from tedious and repetitive tasks but also enhances the speed, quality, and repeatability of the work performed. |
| | Connectivity | Connectivity defines the state of mutual connection between devices, machines, and computer systems, enabling communication and data exchange among the resources of the enterprise. Digitization results in an increasing number of devices connected via cables and systems being transformed from analog forms into wireless and digital equivalents. The number of devices capable of operating within the digital ecosystem of the Internet of Things is growing, leading to a tremendous increase in data generated by them. Simultaneously, the current level of advancement in cloud computing technology and wireless infrastructure allows data to be collected, managed, and analyzed within an integrated system. This enables the standardization of various systems, such as production systems, building infrastructure management, and overall business administration and management. |
| | Autonomization | Autonomization is a kind of ‚intelligence' within the framework of the Fourth Industrial Revolution manufacturing plants, providing the capability for autonomous decision-making within a distributed network of machines and devices. Automation and Connectivity focus on establishing connections between equipment, machines, and IT systems for data collection and integration. Autonomization, on the other hand, involves intelligent processing and analysis of this data, generating the basis for making predictive or prescriptive decisions within the system. This enables the acceleration and flexibility of manufacturing systems, while simultaneously reducing costs through data analysis and autonomization. |

*Contd.*

**Table 24** *Contd.*

| Module | Pillars | Example |
|---|---|---|
| | Intelligent product | The created value according to the concept of Industry 4.0 is primarily the pursuit of new quality that meets individual customer requirements. Companies can achieve this by creating new business models and developing intelligent products. These products embody the ultimate value derived from the vision of Industry 4.0. The latest innovations in design technologies and sensor development allow for the creation of a digital representation of physical products (known as a digital twin) and simulation, monitoring, control, and modification of their parameters. Intelligent products enable the collection of data from the product's usage stage by the customer in real time. This opens up unique opportunities for their utilization by the company, such as creating a complementary offering of services and products. An intelligent product has additional functionality in the form of integrated services, which can significantly change the way the sold value is accounted for, for example, within the as-a-Service model. Additionally, an intelligent product has the capability to make autonomous decisions, thereby deepening the relationship between the company and the customer. |
| **Processes** | Integration with environment | As part of integration with the environment (known as horizontal integration of the value chain), the model distinguishes two perspectives: from the standpoint of the supply chain and customer relationships. The supply chain encompasses planning processes, material and inventory management from the very beginning of the delivery cycle until consumption. The vision of the supply chain in Industry 4.0 creates a digital, transparent, and integrated chain, where processes are connected through sensor networks and managed in an optimized manner using real-time artificial intelligence algorithms. The digitization of the supply chain enables decision-making within the enterprise (e.g., regarding costs, inventory) by considering data from the entire chain rather than individually and temporarily in an isolated manner from the entire supply chain process. Such evolution brings benefits to all stakeholders in the value chain, ensuring, among other things, faster product market entry through reduced lead time, greater flexibility through real-time optimization tailored to changing needs, and increased efficiency and transparency of data in the supply chain. |

*Contd.*

**Table 24** *Contd.*

| Module | Pillars | Example |
|---|---|---|
| | Product lifecycle | The Product Life Cycle refers to the stages that every product goes through—from design, manufacturing, consumption, and service to the final phase of market exit. In the era of the Fourth Industrial Revolution, product lifecycles are significantly shortened, necessitating greater integration and digitization at all stages of the product life cycle, particularly with the simultaneous increase in the trend of product personalization. Industry 4.0 also introduces the concept of the "digital twin" as a virtual representation of digital products, devices, processes, and systems involved in the entire product lifecycle. The digital twin brings many new possibilities and benefits to the company. Firstly, its application in the purchase stage of an element for a more complex system enables virtual testing through simulating its functionality even before the final purchase of the physical product. It replaces the physical prototype, allowing for virtual testing in a fast, scalable, and much more cost-effective manner. This enables the shortening of the design cycle and a quicker response to customer needs. |
| | Internal integration | The internal integration module is directly related to the production process within the company. It involves planning and executing processes leading to the production of goods and services. One of the goals guiding production is cost optimization, which in Industry 4.0 and personalized production is achievable through the application of new digital technologies and new ways of organizing processes using data integration systems. For example, companies can leverage Internet of Things solutions for remote monitoring and decentralized control of assets |
| | Standarization | In the area of energy efficiency management processes, the leading international standard is ISO 50001. This standard encompasses best practices in the creation, implementation, maintenance, and development of energy management systems. The significance of this area is heightened by forecasts of increasing $CO_2$ emission limits for Poland, leading to a rise in electricity production costs. Managing energy efficiency is cited as one of the main challenges in preparing a company for Industry 4.0. It is also an area where digital technologies play a crucial role in identifying areas for improvement and significantly optimizing energy consumption. The second process highlighted as significant in the Standardization Module is the "Purchase of Technology, Machinery, and Equipment" process. The implementation of production technologies (including VR/AR, intelligent mobile robots, cobots, simulation tools, ML algorithms, IoT, cybersecurity systems, predictive |

*Contd.*

**Table 24** *Contd.*

| Module | Pillars | Example |
|---|---|---|
| | | analytics systems, etc.) will become commonplace in companies enhancing their maturity in the field of Industry 4.0. The maturity of this process requires consideration of various investment parameters during the purchasing process, such as the planned lifecycle of the installation, system/technology openness to integration, purchase and system development costs, the maturity of the ordered technology, and ease of technology integration into existing production environments.<br>Multi-criteria analysis of needs and the establishment of assessment criteria based not only on purchase price are good practices in organizing this process. These practices enhance maturity by accelerating the technology implementation process, preparing production systems for scaling, and reducing the total cost of ownership (TCO) of technological installations in the long-term perspective. |
| **Organization** | Strategy | For the change process to proceed smoothly, it is necessary to develop a digitalization strategy for the company that defines directions, key objectives, and success metrics. All company activities are then synchronized and carried out according to a defined plan. The strategy specifies, among other things, how to leverage existing resources and capabilities of the enterprise through their optimization and integration with digital solutions to achieve a competitive advantage. The formal specification of the company's strategy also allows for scaling its goals within the enterprise and clear communication of introduced changes among employees. |
| | Collaboration and paradigms | The organizational structure is its system of principles that define how roles and scope of responsibilities are assigned, monitored, and coordinated. The structure influences how teams behave, interact, and carry out activities resulting from the company's strategy. In the context of Industry 4.0, organizations are introducing greater decentralization of decision-making, delegating decision-making deeper into the organization, fostering more openness to information sharing, and encouraging increased collaboration between teams—both internally and with external partners. Over the long term, this enables companies to make decisions in a more flexible manner, allowing them to respond more quickly to market changes and customer needs. |
| | People | The company is made up of people, and their competencies directly influence the competitive advantage gained. Therefore, talent development is a key factor for success in Industry 4.0. Building competent and flexible teams with a focus on continuous learning and development at every level becomes a critical factor. Management |

*Contd.*

| Module | Pillars | Example |
|---|---|---|
| | | must implement systems and practices that enable people to acquire information and skills about the latest trends and technologies in Industry 4.0. On the other hand, expectations for employees are changing; they need to be increasingly interdisciplinary, quickly adapt to changes, possess communication skills, and have a 'win-win' mindset. This allows formal talent development programs to not only align with the company's business goals but also contribute to creating a culture of self-learning, knowledge sharing, and personal development. |
| | Leadership | The main responsibility of the company's management is to initiate changes and create a vision for the digitization of the enterprise, as well as to motivate people to collaborate, leading to the achievement of a common goal. Strong leadership, supported by a clear and understandable strategy, along with established metrics for effective actions, are crucial for the organization to successfully operate in a dynamic and highly networked business environment. |

*Source:* Own elaboration based on: (Delab, 2020; Thomas and Saleeshya, 2023; Senna et al., 2023; Şengül and Şengül, 2023; Sommer and Proof, 2023; Santiago and Silva, 2023).

is not necessarily a company that uses the most tools or the latest ones. It is an organization that consciously, with appropriately broad knowledge, selects tools that best fit its profile. Determining the maturity of a company is highly dependent on the industry in which a given manufacturing corporation operates, as different areas will require digitization. A common element is ensuring the proper flow of information within the key area of the company. The success of the organization's digital transformation will depend on a series of good decisions made continuously, along with changes in the environment. With one strategic assumption that we want to modernize our methods and processes, we are aware that a digital revolution is taking place (Rabe and Kilic, 2024).

The adopted methodology and definition of six maturity levels enable a precise assessment of maturity along with a description of the level attained by the respondent and the presentation of recommendations for further actions. Also important is the limitation of the number of questions in the model to maintain the respondent's attention and facilitate understanding of the scope of changes that companies face in the context of Industry 4.0 (Delab, 2020).

The digital MM in Industry 4.0 consists of three fundamental modules crucial for the development of Industry 4.0 in manufacturing companies (Thomas and Saleeshya, 2023). These modules are Technologies, Processes, and Organization. Companies should consider all three areas to fully harness the potential associated with Industry 4.0. The pillars are based on 12 key modules representing critical aspects on which companies should focus to become future-ready organizations within the reference model of Industry 4.0. Table 24 describes the particular pillars of this model.

The adopted assessment methodology is primarily focused on the functional purpose of the designed tool, which is the preliminary assessment of the stage of a company's development towards Industry 4.0 represented by the maturity level (Şengül and Şengül, 2023). It also involves providing practical recommendations based on the obtained assessment.

## References


Ahmad, T., Van Looy, A. and Shafagatova, A. 2024. Business Process Performance: Investigating the Impact of Process-oriented Appraisals and Rewards on Success. *Business and Information Systems Engineering,* *66*(1): 67–84.

Alabbadi, S., Seva-Larrosa, P. and García-Lillo, F. 2023. The mediating effect of information technology business strategic maturity on the relationship between organizational behavior and firm performance: A study of the Jordanian maritime industry. *Problems and Perspectives in Management,* *21*(3): 753–763.

Al-Matari, O.M.M., Helal, I.M.A., Mazen, S.A. and Elhennawy, S. 2021. Adopting security maturity model to the organizations' capability model. *Egyptian Informatics Journal,* *22*(2): 193–199.

Alshammari, F.H. 2023. Design of capability maturity model integration with cybersecurity risk severity complex prediction using Bayesian-based machine learning models. *Service Oriented Computing and Applications,* *17*(1): 59–72.

Amarasekara, D.S., Rathnadiwakara, H., Ratnayake, K., Singh, V.P. and Poosala, S. 2022. The Capability Maturity Model as a Measure of Culture of Care in Laboratory Animal Science. *Alternatives to Laboratory Animals,* *50*(6): 437–446.

Bajic, B., Moraca, S. and Rikalovic, A. 2023. Fuzzy maturity model for Smart Manufacturing Readiness: Industry 5.0 perspective. *In: 2023 IEEE Zooming Innovation in Consumer Technologies Conference, ZINC*: 142–147.

Becker J., Knackstedt, R. and Pöppelbuß, J. 2009. Developing maturity models for IT management. *Bus. Inf. Syst. Eng., 1*(3): 213–222.

Blue People. 2023. *Understanding Capability Maturity Model Integration 2.0.* https://www.linkedin.com/pulse/understanding-capability-maturity-model-integration, (Access data: 23.01.2024).

Bitkowska, A. 2009. *Zarządzanie procesami biznesowymi w przedsiębiorstwie.* Wydawnictwo Vizja Press and IT.

Boretti, A. 2023. Technology readiness level of photo-electro-chemical hydrogen production by parabolic dish solar concentrator. *International Journal of Hydrogen Energy, 48*(90): 35005–35010.

Brand, V., Razavian, M. and Ozkan, B. 2023. Development of a Capability Maturity Model for Organization-Wide Green IT Adoption *Lecture Notes in Business Information Processing, 483 LNBIP*: 163–179.

Čengija, S., Kapitan and K.L. 2008. CMMI – Capability Maturity Model Integration. *MIPRO 2008 – 31st International Convention Proceedings: Digital Economy – 5th ALADIN, Information Systems Security, Business Intelligence Systems, Local Government and Student Papers, 5*: 229–234.

Chen, C.-T., Ova, A. and Hung, W.-Z. 2022. A Digital Capability Maturity Model Based on the Hesitant Fuzzy Linguistic Variables. *Lecture Notes in Networks and Systems, 307*: 687–695.

Chockalingam, S., Nystad, E. and Esnoul, C. 2023. Capability Maturity Models for Targeted Cyber Security Training. *Lecture Notes in Computer Science* (including subseries Lecture Notes in Artificial Intelligence and Lecture Notes in Bioinformatics), *14045 LNCS*: 576–590.

Correia, A.M.M., Veiga, C.P.D., Senff, C.O. and Duclós, L.C. 2021. Analysis of the Maturity Level of Business Processes for Science and Technology Parks. *SAGE Open, 11*(3).

Crosby P. 1979. *Quality is Free.* McGraw Hill, New York.

Dal Forno, A.J., Bataglini, W.V., Steffens, F. and Ulson de Souza, A.A. 2023. Maturity model toll to diagnose Industry 4.0 in the clothing industry. *Journal of Fashion Marketing and Management, 27*(2): 201–219.

Das, P., Perera, S., Senaratne, S. and Osei-Kyei, R. 2023. A smart modern construction enterprise maturity model for business scenarios leading to Industry 4.0. *Smart and Sustainable Built Environment, 6*: 45–75.

de Bruin, T., Rosemann, M., Freeze, R. and Kulkarni, U. 2005. Understanding the Main Phases of Developing a Maturity Assessment Model. *Proceedings of the Australasian Conference on Information Systems*, Sydney.

Delab. 2020. *Wsparcie dla Przemysłu 4.0 w Polsce.* https://www.delab.uw.edu.pl/wp-content/uploads/2020/11/przemysl4.0_Opracowanie_DELabUW.pdf., (Access data: 23.01.2024).

Depaoli, P. and Za, S. 2013. Towards the redesign of e-business maturity models for SMEs. *Designing Organizational Systems: An Interdisciplinary Discourse*, 285–300.

Devi, E.T., Wibisono, D. and Mulyono, N.B. 2022. Identifying Critical Capabilities for Improving the Maturity Level of Digital Services Creation Process. *Journal of Industrial Engineering and Management, 15*(3): 498–519.

Dewi, F. and Mahendrawathi, E.R. 2019. Business process maturity level of MSMEs in East Java, Indonesia. *Procedia Computer Science, 161*: 1098–1105.

Doss, D.A., Henley, R., McElreath, D.H., Hong, Q. and Taylor, L.N. 2021. The capability maturity model integrated as a market engineering maturity model. *International Journal of Service Science, Management, Engineering, and Technology, 12*(3): 175–196.

Duan, F. 2023. Analysis of enterprise accounting data management based on maturity model. *Applied Mathematics and Nonlinear Sciences, 9*(1).

Eby, K. 2022. *Essential Guide to Process Maturity: Models, Tips, and Templates.* https://www.smartsheet.com/content/process-maturity (Access data: 23.01.2024).

Ehrensperger, R., Sauerwein, C. and Breu, R. 2023. A Maturity Model for Digital Business Ecosystems from an IT Perspective. *Journal of Universal Computer Science, 29*(1): 34–72.

El Chaabi, A., Anevlavis, N. and Thuesen, C. 2023. Conceptual Maturity Model for Strong Owner Capability Development in the Construction Industry. *Springer Proceedings in Business and Economics*, 313–325.

Englezos, P. 2023. Technology readiness level of gas hydrate technologies. *Canadian Journal of Chemical Engineering, 101*(6): 3034–3043.

Erhard, A., Arthofer, K. and Helm, E. 2023. Extending a Data Management Maturity Model for Process Mining in Healthcare. *Studies in Health Technology and Informatics, 301*: 192–197.

Fukas, P., Rebstadt, J., Remark, F. and Thomas, O. 2021. Developing an artificial intelligence maturity model for auditing, *133. In: 29th International Conference on Information Systems (ECIS) 2021 Research Papers*.

Gałuszka, M. 2011. Modele dojrzałości biznesowych – analiza porównawcza. *Management Sciences, 8*: 66–74.

Ganesh, H. 2023. *5 Stages of Industry 4.0 Maturity Model.* https://www.fogwing.io/industry-4-0/ industry-40-maturity-model/, (Access data: 23.01.2024).

Gao, Z., Xing, F. and Peng, G. 2023. Research on the Capability Maturity Model of Data Security in the Era of Digital Transformation. *Lecture Notes in Computer Science (including subseries Lecture Notes in Artificial Intelligence and Lecture Notes in Bioinformatics). 14045 LNCS*: 151–162.

Garbin, F.G.B., ten Caten, C.S. and Jesus Pacheco, D.A. 2022. A capability maturity model for assessment of active learning in higher education. *Journal of Applied Research in Higher Education, 14*(1): 295–316.

Glavan, L.M. 2019. Determining business process maturity levels by using cluster analysis: Case of Croatia. *Proceedings of the 15th International Symposium on Operational Research, SOR 2019*: 308–313.

Glaveli, N., Alexiou, M., Maragos, A., Daskalopoulou, A. and Voulgari, V. 2023. Assessing the Maturity of Sustainable Business Model and Strategy Reporting under the CSRD Shadow. *Journal of Risk and Financial Management, 16*(10): 445.

Glogovac, M., Ruso, J. and Maricic, M. 2022. ISO 9004 maturity model for quality in industry 4.0. *Total Quality Management and Business Excellence, 33*(5–6): 529–547.

Gracey, A. 2020. Building an organizational resilience maturity framework. *Journal of Business Continuity and Emergency Planning, 13*(4): 313–327.

Hammer M. 2007. *The Process Audit.* http://www.krajciova.sk/Knihy/BPR/Michael%20Hammer%20 -%20The%20Process%20Audit%20-%200407.pdf, (Access data: 23.01.2024).

Harrington, H.J. 2006. *Process Management Excellence. The Art of Excelling in Process Management.* Paton Press, California 2.

Hassan, N.H. and Arshad, N.I. 2023. Information technology personnel competency towards organizational agility: Study at Malaysia automotive. *Indonesian Journal of Electrical Engineering and Computer Science, 32*(1): 312–317.

Hein-Pensel, F., Winkler, H., Brückner, A., Friedrich, J. and Zinke-Wehlmann, C. 2023. Maturity assessment for Industry 5.0: A review of existing maturity models. *Journal of Manufacturing Systems, 66*: 200–210.

Henriquez, R., Muñoz-Villamizar, A. and Santos, J. 2023. Key factors in operational excellence for Industry 4.0: An empirical study and maturity model in emerging countries. *Journal of Manufacturing Technology Management, 34*(5): 771–792.

Humphrey W.S. 1989. *Managing the Software Process.* Addison-Wesley Professional, New York.

Hyrynsalmi, S., Olsson, H.H. and Bosch, J. 2023. Towards a Data Business Maturity Model for Software-intensive Embedded System Companies. *Proceedings of the 29th International Conference on Engineering, Technology, and Innovation: Shaping the Future, (ICE) 2023*.

ISO 10014:2021. *Quality Management Systems. Managing an Organization for Quality Results. Guidance for Realizing Financial and Economic Benefits.* ISO International Organization for Standardization.

ISO 9001:2015. *Quality Management Systems. Requirements.* ISO International Organization for Standardization.

ISO 9004:2018. *Quality Management Quality of an Organization. Guidance to Achieve Sustained Success*. ISO International Organization for Standardization.

Istaitih, J. 2020. *Industry 4.0 Maturity Index*. https://www.linkedin.com/pulse/industry-40-maturity-index-jameel-istaitih-%D7%92-%D7%9E%D7%99%D7%9C-%D7%90%D7%A1%D7%AA%D7%99%D7%AA%D7%94/?articleId=6627486892382277633%C2%A0, (Access data: 23.01.2024).

Jesus, C., Lima, R.M., Barretiri, L. and Lopes, S.I. 2022 (June). A Maturity Model Proposal with Readiness Level Assessment for Organizational Processes Improvement. *Iberian Conference on Information Systems and Technologies, CISTI*.

Jordan, S.W., Ivey, S., Levy, M., Palazolo, P. and Waldron, B. 2022. Complete Streets: A New Capability Maturity Model. *Journal of Urban Planning and Development, 148*(1): 04021071.

Kalinowski, T.B. 2011. Modele oceny dojrzałości procesów. *Acta Universitatis Lodziensis, Folia Oeconomica, 258*: 173–187.

Kenneally, J., Curley, M., Wilson, B. and Porter, M. 2013. Enhancing Benefits from Healthcare IT Adoption using Design Science Research: Presenting a Unified Application of the IT Capability Maturity Framework and the Electronic Medical Record Adoption Model. *Communications in Computer and Information Science, 388 CCIS*: 124–143.

Khraiwesh, M. 2020. Measures of organizational training in the capability maturity model integration (CMMI). *International Journal of Advanced Computer Science and Applications, 2*: 584–592.

Kimita, K., McAloone, T.C., Ogata, K. and Pigosso, D.C.A. 2022. Servitization maturity model: Developing distinctive capabilities for successful servitization in manufacturing companies. *Journal of Manufacturing Technology Management, 33*(9): 61–87.

Kozlov, A.V., Zaychenko, I.M. and Kolotova, D.P. 2023. Business Digital Maturity Assessment in Strategic Decision Making. *Lecture Notes in Networks and Systems, 684 LNNS*: 921–934.

Lavin, A., Gilligan-Lee, C.M., Visnjic, A., Parr, J. and Gal, Y. 2022. Technology readiness levels for machine learning systems. *Nature Communications, 13*(1): 6039.

Leyh, C., Bley, K., Schaffer, T. and Forstenhausler, S. 2016. SIMMI 4.0-a maturity model for classifying the enterprise-wide it and software landscape focusing on Industry 4.0. *Proceedings of the 2016 Federated Conference on Computer Science and Information Systems, FedCSIS*, 1297–1302, 7733413.

Leyh, C. and Schäffer, T. 2018. Simmi 4.0: System integration maturity model industry 4.0. *VDI Berichte, 2018*(2330): 1105–1117.

Leyh, C., Schäffer, T., Bley, K. and Forstenhäusler, S. 2017. Assessing the it and software landscapes of industry 4.0-enterprises: The maturity model SIMMI 4.0. *Lecture Notes in Business Information Processing, 277*: 103–119.

Limat, C. 2022. Disruptionspotenzial künstlicher Intelligenz: Ein Reifegradmodell zur Einführung ganzheitlicher KI-Initiativen in Unternehmen. *Wirtschaftsinformatik Manage, 14*(1): 60–67.

Liu, G.-S., Zheng, T.-T., Yi, H.-G. and Ding, T.-X. 2023. Cloud Knowledge Capability Maturity Model Integration and Evaluation Method Using a Cloud Knowledge Management System. *IEEE Engineering Management Review, 51*(3): 93–108.

Maris, A., Ongena, G. and Ravesteijn, P. 2023. Business Process Management Maturity and Process Performance: A Longitudinal Study. *Lecture Notes in Business Information Processing, 490 LNBIP*: 355–371.

Martins, G.R.D.N., Ramos, L.F.P., Loures, E.F.R., do Amaral, L.M.B. and Deschamps, F. 2023. Prioritization of Industry 4.0 Technologies to Increase Maturity in Lean Manufacturing from the Perspective of Enterprise Interoperability. *Lecture Notes in Mechanical Engineering*, 379–386.

Mazur, A. 2014. Self-assessment of Maturity of Organization in Terms of Occupational Health and Safety with the Recommendations of ISO 9004:2010. *Communications in Computer and Information Science, 435 PART II*: 479–484.

Merdin, D., Ersoz, F. and Taskin, H. 2023. Digital Transformation: Digital Maturity Model for Turkish Businesses. *Gazi University Journal of Science, 36*(1): 263–282.

Mittal, S., Khan, M.A., Romero, D. and Wuest, T. 2018. A critical review of smart manufacturing and Industry 4.0 maturity models: Implications for small and medium-sized enterprises (SMEs). *Journal of Manufacturing Systems, 49*: 194–214.

Moradi, T., Jafari, M., Maleki, M.R., Naghdi, S. and Ghiasvand, H. 2015. Quality Management Systems Implementation Compared with Organizational Maturity in Hospital. *Global Journal of Health Science*, *8*(3): 174–182.

Moradi-Moghadam, M., Safari, H. and Maleki, M. 2013. A novel model for business process maturity assessment through combining maturity models with EFQM and ISO 9004:2009. *International Journal of Business Process Integration and Management*, *6*(2): 167–184.

North, N. and Coetzee, M. 2022. Development of a capability maturity model for the establishment of children's nursing training programs in southern and eastern Africa. *Evaluation and Program Planning*, *91*: 102061.

Onyeme, C. and Liyanage, K. 2023. A systematic review of Industry 4.0 maturity models: applicability in the O&G upstream industry. *World Journal of Engineering*, *20*(6): 1160–1173.

Oruthotaarachchi, C.R. and Wijayanayake, W.M.J.I. 2023. Developing a multi-perspective capability model for organizational business process management maturity assessment in digital era. *International Journal of Information Systems and Change Management*, *13*(3): 191–208.

Ouazzani-Chahidi, A., Abdellatif, L., Jimenez, J.-F. and Berrah, L. 2023. Maturity levels of management process for improving industrial performance. *Scientific African*, *21*: e01852.

Ozkan, B., Koops, M., Türetken, O. and Reijers, H.A. 2023. The Influence of Business Process Management System Implementation on an Organization's Process Orientation: A Case Study of a Financial Service Provider. *Information Systems Management*, *5*: 59–83.

Pane, E.S. and Sarno, R. 2015. Capability Maturity Model Integration (CMMI) for Optimizing Object-Oriented Analysis and Design (OOAD). *Procedia Computer Science*, *72*: 40–48.

Peng, L., Feng, W., Chen, K. and Li, C. 2016. Smart manufacturing capability maturity model: Connotation, feature and trend. *Proceedings of the International Conference on Electronic Business (ICEB)*, 151–157.

Pöppelbuß, J. and Röglinger, J. 2011. What makes a useful maturity model? A framework of general design principles for maturity models and its demonstration in business process management. *In*: *ECIS 2011 Proceedings. AIS Electronic Library (AISeL)/Association for Information Systems*.

Rabe, M. and Kilic, E. 2024. Framing the Digital Business Process Twin: From a Holistic Maturity Model to a Specific and Substantial Use Case in the Automotive Industry. *Lecture Notes in Business Information Processing*, *492 LNBIP*: 353–364.

Riaz, M.N. 2017 (December). Factors affecting the transition time between capability maturity model integration (CMMI) levels in software industry of Pakistan: An empirical study. *2017 International Conference on Information and Communication Technologies, ICICT 2017*, 90–96.

Richter, T., Schmidt, D., Hahlweg, H., Behdinan, K. and Albers, A. 2020. Objective-based process model for enhancing the product maturity level in the early phase of a development process. *Procedia CIRP*, *91*: 207–213.

Saad, S.M., Bahadori, R. and Jafarnejad, H. 2021. The smart SME technology readiness assessment methodology in the context of industry 4.0. *Journal of Manufacturing Technology Management*, *32*(5): 1037–1065.

Salazar, O. 2022. *What is the Technology Maturity Scale (TRL)?* https://euro-funding.com/en/blog/what-is-the-technology-maturity-scale-trl/, (Access data: 23.01.2024).

Sansalvador, M.E. and Brotons, J.M. 2017. The application of Owas in expertise processes: The development of a model for the quantification of hidden quality costs. *Economic Computation and Economic Cybernetics Studies and Research*, *51*(3): 73–90.

Sansalvador, M.E. and Brotons, J.M. 2023. A fuzzy dynamic model for total quality cost. *Economic Computation and Economic Cybernetics Studies and Research*, *57*(2): 5–20.

Santacruz, R.F.B., Sullivan, B.P., Terzi, S. and Sassanelli, C. 2023. Developing a Technology Readiness Level Template for Model-Based Design Methods and Tools in a Collaborative Environment. *IFIP Advances in Information and Communication Technology*, *667 IFIP*: 237–249.

Santiago, S.B. and Silva, J.R. 2023. Strategic Roadmap for Digital Transformation Based on Measuring Industry 4.0 Maturity and Readiness. *Communications in Computer and Information Science*, *1886 CCIS*: 336–347.

Santos, B.M.O., Dias, F.J.M., Trillaud, F., Sotelo, G.G. and de Andrade Junior, R. 2023. A Review of Technology Readiness Levels for Superconducting Electric Machinery. *Energies, 16*(16): 5955.

Schmele, J.A. and Foss, S.J. 1989. The quality management maturity grid: A diagnostic method. *Journal of Nursing Administration, 19*(9): 29–34.

Schumacher, A., Erol, S. and Sihn, W. 2016. A maturity model for assessing industry 4.0 readiness and maturity of manufacturing enterprises. *Procedia CIRP, 52*: 161–166.

Schumacher, A., Nemeth, T. and Sihn, W. 2019. Road mapping towards industrial digitalization based on an Industry 4.0 maturity model for manufacturing enterprises. *Procedia CIRP, 79*: 409–414.

Seipolt, A., Buschermohle, R., Hofinghoff, M., Korn, G.-H. and Schumacher, M. 2023. Technology readiness levels of reinforcement learning methods for simulation-based production scheduling. *Lecture Notes in Informatics (LNI), Proceedings–Series of the Gesellschaft fur Informatik (GI), P-337*: 1377–1392.

Semeraro, C., Alyousuf, N., Kedir, N.I. and Lail, E.A. 2023. A maturity model for evaluating the impact of Industry 4.0 technologies and principles in SMEs. *Manufacturing Letters, 37*: 61–65.

Şengül, U. and Şengül, A.B. 2023. Digital maturity levels in transition to industry 4.0 in businesses: A qualitative analysis and fuzzy MCDM applications in machinery manufacturing industries. *Management in the Digital Era: Different Perspectives*, 173–189.

Senna, P., Barros, A.C., Bonnin Roca, J. and Azevedo, A. 2023. Development of a digital maturity model for Industry 4.0 based on the technology-organization-environment framework. *Computers and Industrial Engineering, 185*: 109645.

Shen, L., Du, X., Cheng, G. and Wei, X. 2021. Capability Maturity Model (CMM) method for assessing the performance of low-carbon city practice. *Environmental Impact Assessment Review, 87*: 106549.

Shmeleva, N., Tolstykh, T. and Dudareva, O. 2023. Integration as a Driver of Enterprise Sustainability: The Russian Experience. *Sustainability, 15*(12): 9606.

Simetinger, F. and Basl, J. 2022. A pilot study: An assessment of manufacturing SMEs using a new Industry 4.0 Maturity Model for Manufacturing Small- and Middle-sized Enterprises (I4MMSME). *Procedia Computer Science, 200*: 1068–1077.

Siuta, D., Kukfisz, B., Kuczyńska, A. and Mitkowski, P.T. 2022. Methodology for the Determination of a Process Safety Culture Index and Safety Culture Maturity Level in Industries. *International Journal of Environmental Research and Public Health, 19*(5): 2668.

Skalli, D., Charkaoui, A. and Cherrafi, A. 2023. Design of Industry 4.0 Maturity Model and Readiness Assessment Under the Digital Transformation towards the Implementation of LSS4.0. *Lecture Notes in Networks and Systems, 771 LNNS*: 167–180.

Skyttermoen, T. and Wedum, G. 2023 (November). Developing Capabilities for Sustainable Business Models: Exploring Project Maturity for Innovation Processes. *Proceedings of the European Conference on Management, Leadership, and Governance, 2023,-* 370–379.

Slack N. 2007. Chambers, S. and Johnston, R. 2007. *Operations Management*. Pearson Education, Essex.

Sliż, P. 2021. Identification of factors influencing the level of maturity of dealerships in Norway process orientation. *Knowledge and Process Management, 28*(4): 353–363.

Sommer, S. and Proff, H. 2023. Digital Maturity in the German Traditional Industries: Status Quo, Profit Impact, and Paths of Acceleration. *International Journal of Innovation and Technology Management, 20*(4): 2350019.

Song, J., He, Z., Jiang, L., Liu, Z. and Leng, X. 2023. Synergy Management of a Complex Industrial Production System from the Perspective of Flow Structure. *Systems, 11*(9): 453.

Souhail, S., Ibtissam, E.H. and Anass, C. 2023. Maturity Models as a Support for Industry 4.0 Implementation: Literature Review. *Lecture Notes in Networks and Systems, 771 LNNS*: 123–136.

Sukrat, S. and Leeraphong, A. 2024. A digital business transformation maturity model for micro enterprises in developing countries. *Global Business and Organizational Excellence, 43*(2): 149–175.

Tarí J.J. 2010. Self-assessment processes: The importance of follow-up for success. *Quality Assurance in Education, 1*: 522–538.

The Industry 4.0 Maturity Index: 6 Levels Toward Digital Transformation, https://matics.live/blog/industry-4-0-maturity-index/ (Access data: 23.01.2024).

Thomas, T. and Saleeshya, P.G. 2023. CMMI-based fuzzy logic approach to assess the digital manufacturing maturity level of manufacturing industries. *TQM Journal, 35*(8): 2658–2683.

Treviño-Elizondo, B.L. and García-Reyes, H. 2023. An Employee Competency Development Maturity Model for Industry 4.0 Adoption. *Sustainability, 15*(14): 11371.

Tripathi, A., Nassereddine, H., Sturgill, R.E., Hatoum, M.B. and Ammar, A. 2024. People, Process, and Technology Maturity Levels for Successful Technology Implementation by State Departments of Transportation. *Transportation Research Record, 2678*(1): 12–21.

Van Looy, A. 2014. Business Process Maturity: A Comparative Study on a Sample of Business Process Maturity Models. Springer International Publishing, London.

Van Looy, A. 2018. On the synergies between business process management and digital innovation. *Lecture Notes in Computer Science (including subseries Lecture Notes in Artificial Intelligence and Lecture Notes in Bioinformatics), 11080 LNCS*, 359–375.

Wang, W., Wang, J., Chen, C., ...Chu, C. and Chen, G. 2022. A Capability Maturity Model for Intelligent Manufacturing in Chair Industry Enterprises. *Processes, 10*(6): 1180.

Wolniak, R. 2013. The assessment of significance of benefits gained from the improvement of quality management systems in Polish organizations. *Quality and Quantity, 47*(1): 515–528.

Wood, P.B. and Vickers, D. 2018 (March) Anticipated impact of the capability maturity model integration (CMMI®) V2.0 on aerospace systems safety and security. *IEEE Aerospace Conference Proceedings, 2018*, 1–11.

Zhai, Q., Li, H., Lv, K. and Qin, Y. 2022. Research on the Development Path of Cost Consulting Enterprises under Whole-Process Engineering Consultancy. Carbon Peak and Neutrality Strategies of the Construction Industry. *Proceedings of the 2022 International Conference on Construction and Real Estate Management, 5*: 792–799.

# 3 | Measuring the Maturity of Companies

Chapter 3 presents the concept of measuring the business maturity of enterprises. The chapter begins with the conceptualization and operationalization of a research model aimed at assessing the business maturity of industrial enterprises. In the conceptual phase, the dimensions of business maturity were indicated and a research model consistent with the digital maturity model of Industry 4.0 was defined. In the operational phase, a digital maturity measurement model was defined covering nine pillars of Industry 4.0: Internet of Things, Big Data, Cloud Computing, Advanced Simulation, Autonomous Systems, Universal Integration, Virtual and Augmented Reality, Additive Manufacturing, and Cybersecurity. The chapter presents the methodological approach, including the construction of a business maturity measurement tool and the determination of the reliability of the proposed measurement scale. The results, analyzed using statistical programs indicate high reliability of the measurement scale. The research process is described in seven stages, from model creation to data analysis. The findings aim to gain insight into the technological advancement and business maturity of Polish manufacturing enterprises.

## 1. Conceptualization and Operationalization of the Research Model for Business Maturity of Polish Manufacturing Companies

The conceptualization and operationalization of the research model serve as a starting point for subsequent research tasks, which will include organizing the methods of data collection necessary to achieve the established research objectives, appropriately preparing the data for analysis, and analyzing the obtained results.

The researched issue concerning the level of business maturity of industrial enterprises is relatively new and aims to explain the essence of the phenomenon and the relationships between the level of business maturity of enterprises and the effectiveness of these enterprises in implementing the planned strategy. The purpose of the study is to determine the nature of the phenomenon, its structure, and the methods of measurement. Thus, the goal of the research (and this chapter) is to define a conceptual model of business maturity for enterprises, propose its measurement model, and, in the context of the proposed solutions, demonstrate that it can serve as an additional dimension in the comparative analysis of sectors and industries in Poland.

The specific objectives are formulated as follows:

1. Organize the concepts, definitions, and classifications related to the business maturity of industrial enterprises.
2. Formulate a conceptual research model of the business maturity of industrial enterprises.
3. Operationalize the conceptual model of the business maturity of industrial enterprises.
4. Measure the individual elements of the business maturity model of enterprises according to the research model and assess the validity and reliability of the measurement.

The process of constructing the research conceptual model was preceded by a literature review in the area of industrial transformation (Chapter 1) and business maturity models (Chapter 2). The theory of industrial transformation towards Industry 4.0 was identified and described, indicating the paths leading to achieving business maturity of enterprises. Business maturity models were described and classified, indicating the components of these models, including the maturity of organizational processes, quality management systems, and technological readiness. The relationships and importance of the level of business maturity of enterprises in building strategies, undertaking cooperation, and achieving leadership in the digital economy were highlighted. A critical analysis of the literature formed the basis for constructing an original research model for business maturity of enterprises.

It is assumed that the business maturity of enterprises is the organization's ability to adapt, change, and develop in a dynamic market environment. It is the ability to improve the processes, structures, and resources of the enterprise to achieve specific strategic and operational goals. It indicates the level of skills and readiness of the enterprise to perform tasks and achieve established objectives. The elements of the definition of business maturity for enterprises are presented in Table 1.

**Table 1** Components of the definition of business maturity of enterprises.

| Business Maturity | |
|---|---|
| Ability to … | adapt |
| | change |
| | develop |

*Contd.*

**Table 1** *Contd.*

| Business Maturity | |
|---|---|
| Capability for enhancement | processes |
| | structures |
| | resources |
| Ready to … | Implement/accomplish tasks |
| | achieve goals |

*Source:* Own elaboration.

The ability to adapt, change, and develop, as well as the capability to improve processes, structures, and resources, and the readiness to accomplish tasks and achieve goals, can be defined in various dimensions. The key dimensions include technology, processes, and organization. Business maturity in the technology dimension encompasses automation, connectivity, autonomy, and smart products. The process dimension of business maturity is defined by processes of integration with the environment, product development, internal integration, and standardization. The organizational dimension of business maturity pertains to strategy, collaboration, people, and leadership (Table 2).

**Table 2**   Dimensions of business maturity of enterprises.

| Dimension | Elements |
|---|---|
| Technologies | Automation |
| | Connectivity |
| | Autonomization |
| | Intelligent product |
| Processes | Integration with environment |
| | Product lifecycle |
| | Internal integration |
| | Standardization |
| Organization | Strategy |
| | Collaboration |
| | People |
| | Leadership |

*Source:* Own elaboration based on: Table 24 (Chapter 2).

The methodology adopted in this study for assessing the level of business maturity of enterprises focuses on the functional objective of the designed research model, which is to evaluate the stage of development of an enterprise towards Industry 4.0. In the research model, the concept of Industry 4.0 served as the primary research paradigm for the authors.

Industry 4.0 maturity is conceptualized as a holistic model that assesses an organization's readiness and advancement in implementing Industry 4.0 technologies. This model considers various dimensions aligned with the so-called pillars of Industry 4.0 to evaluate the technological advancement of an enterprise in the process of transitioning towards an intelligent and interconnected ecosystem, thereby assessing its business maturity in the digital economy.

The Industry 4.0 maturity model serves as a metric for evaluating the effects of transformation on business activities and maturity. It allows for the assessment of enterprise maturity, helping to identify strengths and weaknesses as well as areas requiring improvement on the path to a technologically advanced future, which is a necessary condition for achieving full business maturity. Increasing efficiency, effectiveness, and productivity is not possible without the implementation of digital technologies across multiple dimensions of enterprise activity.

Based on the analysis of the literature, key dimensions identified in Industry 4.0 maturity models were selected for the study: digital infrastructure, data management, analytics, process automation, and system integration (Table 3).

**Table 3**  Dimensions of the Industry 4.0 maturity model used in the study.

| Source | Dimension | Number of Dimension |
|---|---|---|
| Manufacturing Enterprise 4.0 Maturity Model (MEMM) | Digital Infrastructure | D1 |
| | Analytics | D2 |
| | Process Automation | D3 |
| System Integration Maturity Model Industry 4.0 (SIMMI 4.0) | System Integration | D4 |
| | Data Management | D5 |

*Source:* Own elaboration based on: Table 23 (Chapter 2).

The operationalization of the model for measuring the business maturity of enterprises was based on the pillars of Industry 4.0 and selected elements characterizing a each pillar (Table 4 and Figure 1).

**Table 4**  Pillars and elements of Industry 4.0 and the variables assigned to them in the business maturity measurement model of industrial enterprises.

| Pillars of Industry 4.0 | Elements of Pillars | Item in Measurement Model |
|---|---|---|
| Internet of Things P1 | Intelligent communication, visualization, and process response systems | q1 |
| | Technical-business solutions based on the capabilities of the Internet of Things (IoT) | q2 |
| | Radio frequency identification (RFID) and product identification codes | q3 |
| | Direct machine-to-machine (M2M) communication | q4 |

Contd.

**Table 4** *Contd.*

| *Pillars of Industry 4.0* | *Elements of Pillars* | *Item in Measurement Model* |
|---|---|---|
| Big Data P2 | Machine parameters and vehicle locations | q5 |
| | Software for real-time processing and analysis of data on processes/machines/products | q6 |
| | Managerial dashboards | q7 |
| | Key performance indicators (KPIs) monitoring process parameters | q8 |
| | Big Data algorithms (maintenance - PM) | q9 |
| | Data base of customer purchasing | q10 |
| Cloud Computing P3 | Data collected and stored in the cloud | q11 |
| | Data-based services | q12 |
| | Data processed in the cloud | q13 |
| Advanced Simulation P4 | Advanced computer simulations of production processes | q14 |
| | Digital models for prototyping and process improvement | q15 |
| Autonomous Systems P5 | Automated and robotized production lines | q16 |
| | Autonomous IT-computer systems for controlling their devices and responding to commands from external control systems | q17 |
| | Transport (traffic) scenarios with object and vehicle detection projects (using drones for object monitoring) | q18 |
| | Industrial and service robots are utilized in collaboration with humans (operators) | q19 |
| | Autonomous vehicles (AGVs) in internal transport or logistics | q20 |
| Universal Integration P6 | Computer communication and process handling systems with external user systems to streamline communication and document flow | q21 |
| | A flexible production environment allowing real-time production corrections (autonomous robots, autonomous control systems) | q22 |
| Virtual and Augmented Reality P7 | Virtual Reality (VR) techniques | q23 |
| | Augmented Reality (AR) techniques | q24 |

*Contd.*

**Table 4**  *Contd.*

| Pillars of Industry 4.0 | Elements of Pillars | Item in Measurement Model |
|---|---|---|
| Additive Manufacturing P8 | 3D printing for designing solutions and products for customers (commercial product printing) | q25 |
| | 3D printing to ensure equipment continuity (printing spare parts for own use) | q26 |
| | 3D printing to print prototypes (models) to shorten product design time | q27 |
| Cybersecurity P9 | Dedicated cybersecurity systems | q28 |
| | Security measures such as passwords, encryption, and codes to enhance data security | q29 |

*Source:* Own elaboration based on: Table 4 (Chapter 1).

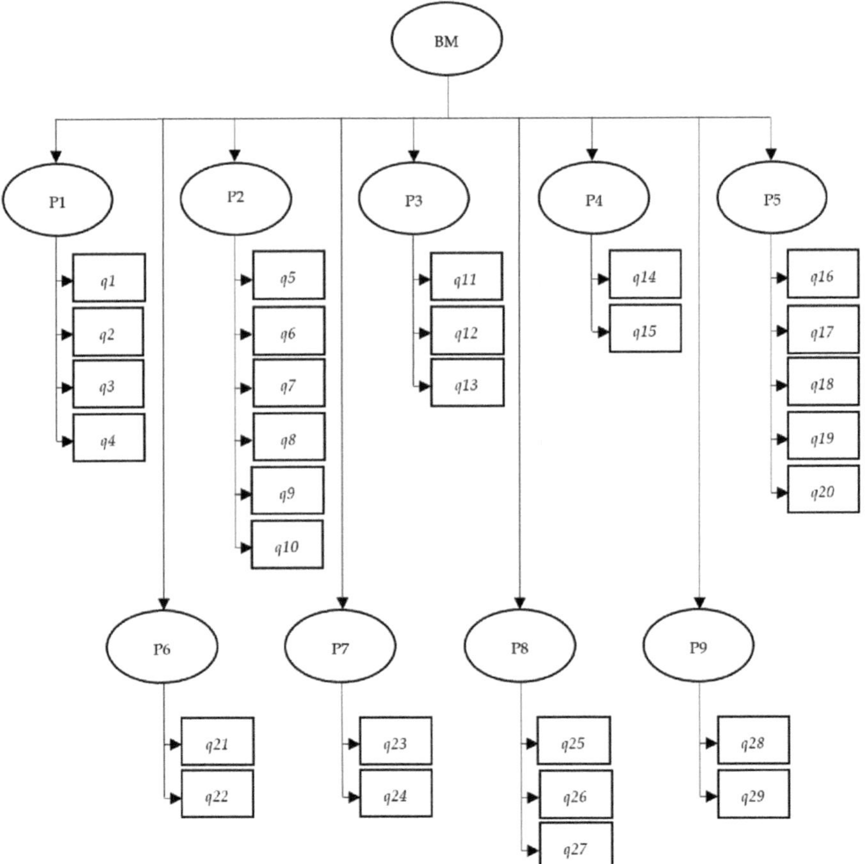


**Fig. 1**  Operationalized measurement model of Business Maturity (BM) of enterprises
*Source:* Own study.

The identified pillars of Industry 4.0 (P1–P9) and selected elements that building these pillars are part of the dimensions of the business maturity model of enterprises. Table 5 compares the items of the measurement model (q1– q29) with the dimensions of the business maturity model of enterprises.

Business maturity of an enterprise is perceived as an incremental process, characterized by various levels of advancement achieved through successive stages. Each stage defines the criteria for a given level of complexity and advancement. At the lowest level of maturity, an enterprise may exhibit basic skills and structures that, however, require significant improvements and optimization. As the organization develops and progresses through the stages of maturity, its processes become increasingly optimized, standardized, and efficient. The highest level of maturity signifies the organization's full capability to perform its tasks efficiently, effectively, and excellently, in accordance with best practices in the field.

For the purpose of measuring the business maturity of manufacturing enterprises, six levels of maturity have been distinguished. On a scale of 0–5, the level of advancement in the use of individual elements of the Industry 4.0 pillars (as listed in Table 4) was measured in manufacturing enterprises.

The measurement scale used in the business maturity assessment model of manufacturing enterprises is designed to determine the degree of implementation of Industry 4.0 solutions in various areas of the organization's activities. This scale includes six levels, with each level characterized by a different degree of advancement in the implementation of technologies and practices associated with Industry 4.0. Level 0 indicates a complete absence of Industry 4.0 practices, where there is no evidence of their presence in the organization. Level 1 indicates that these practices are established but still in the early initiative phase. At Level 2, Industry 4.0 solutions are noticeable in some areas, indicating moderate, isolated use of technology in selected processes, products, or technologies. Level 3 describes a broad but not yet widespread application of these solutions, encompassing about half of the organization's activities. At Level 4, these solutions are very typical, with only a few exceptions, indicating comprehensive use in most processes, products, and technologies. The highest level, 5, signifies full and comprehensive implementation of Industry 4.0 solutions across the entire organization, practically without exceptions, and encompasses all activities in the entire value chain. This scale allows for a detailed assessment of the technological advancement of manufacturing enterprises, identifying areas requiring further development and optimization.

In the next step of the research procedure, the conceptual and operational assumptions of the business maturity measurement model for manufacturing enterprises were, translated into a measurement tool, which was subjected to reliability assessment.

**Table 5** Matrix of measurement model items and enterprise business maturity model dimensions.

| Pillars of Industry 4.0 | Number of Items | Number of BM Dimensions | | | | |
|---|---|---|---|---|---|---|
| | | D1 | D2 | D3 | D4 | D5 |
| **P1** | q1 | | | | ✓ | |
| | q2 | | | | ✓ | |
| | q3 | ✓ | | | | |
| | q4 | | | | ✓ | |
| **P2** | q5 | | ✓ | | | ✓ |
| | q6 | | ✓ | | | |
| | q7 | | | | | ✓ |
| | q8 | | ✓ | | | |
| | q9 | | ✓ | | | |
| | q10 | | | | | ✓ |
| **P3** | q11 | ✓ | | | | |
| | q12 | | | | ✓ | |
| | q13 | | | | | ✓ |
| **P4** | q14 | | | | | ✓ |
| | q15 | | | | ✓ | |
| **P5** | q16 | | | ✓ | | |
| | q17 | | | ✓ | ✓ | |
| | q18 | | | | ✓ | |
| | q19 | | | ✓ | | |
| | q20 | | | ✓ | | |
| **P6** | q21 | | | | ✓ | |
| | q22 | | | | | ✓ |
| **P7** | q23 | ✓ | | | | |
| | q24 | ✓ | | | | |
| **P8** | q25 | | | | ✓ | |
| | q26 | | | | ✓ | |
| | q27 | | | | ✓ | |
| **P9** | q28 | | | | | ✓ |
| | q29 | | | | | ✓ |

**Note:** The symbol ✓ indicates that the item in the measurement model represents a given dimension of the business maturity.

*Source:* Own elaboration.

## 2.  Research Tools: Composition and Descriptive Analysis

Based on the operationalized research model of business maturity of Polish manufacturing companies, a questionnaire for quantitative research was prepared. The questionnaire consisted of the following parts: preamble and invitation to participate in the research, explanation of the measurement scale, main questions, and demographics section.

The main questions were divided according to the nine pillars of the Industry 4.0 concept (Rüßmann et al., 2015; Pollak, 2022) discussed in the first chapter of this monograph. Accordingly, questions were constructed for the following Industry 4.0 pillars: Internet of Things, Big Data, cloud computing, advanced simulation, autonomous systems, universal integration, augmented reality, additive manufacturing/3D printing, and cybersecurity. In total, 29 questions were posed.

The Internet of Things (IoT) is a technological approach where devices are networked to utilize software and automation procedures, enabling intelligent operations. Any device equipped with smart sensor technology can be part of the IoT environment, as well as various objects can be included in this network. The following questions regarding the Internet of Things were posed in the questionnaire:

- *q1. In the enterprise, intelligent communication, visualization, and process response systems are operational (e.g., Andon system/boards),*
- *q2. In the enterprise, technical-business solutions based on the capabilities of the Internet of Things (IoT) and/or industrial platforms have been implemented,*
- *q3. In the enterprise, radio frequency identification (RFID) and product identification codes are used to control supply chain/production/logistics/ warehouse,*
- *q4. In the enterprise, direct machine-to-machine (M2M) communication is facilitated using distributed device sensors and other components of industrial networks.*

Big Data is a highly advanced technology used for collecting, integrating, and analyzing vast amounts of data generated by sensors, products, machines, etc., which supports decision-making, communication, coordination, and innovation implementation. Five questions regarding the use of Big Data in Polish manufacturing companies were posed in the questionnaire:

- *q5. In the enterprise, machine parameters and vehicle locations are tracked to streamline processes,*
- *q6. In the enterprise, software (computer system) is utilized for real-time processing and analysis of data on processes/machines/products,*
- *q7. In the enterprise, managerial dashboards (providing access to current information at the managerial level) are used for processes' control (monitoring) and to make decisions,*
- *q8. In the enterprise, data on Key Performance Indicators (KPIs) monitoring process parameters are collected on a large scale (management system),*

- *q9. In the enterprise, Big Data algorithms are employed for diagnosing production processes and machine self-diagnosis (maintenance - PM),*
- *q10. In the enterprise, large datasets are analyzed to determine customer purchasing preferences and customer profile.*

Cloud computing is a technology that enables the storage and processing of data on remote servers accessible via the internet, allowing users to access resources from anywhere. It facilitates real-time data exchange, enhances agility and collaboration in global enterprise networks, and simplifies the scalability of operations without the need for investment in local computing infrastructure. Three questions were used to measure cloud computing:

- *q11. In the enterprise, data is collected and stored in the cloud,*
- *q12. In the enterprise, access to various data-based services is provided, and these data are made available on cloud platforms,*
- *q13. In the enterprise, data collected and processed in the cloud are used for easy scaling of process support computer systems.*

Advanced simulations are models used for testing and optimizing machines, new products, and processes, as well as predicting problems before they occur, including digital twins, which are virtual replicas of physical objects that allow for the simulation of planned solutions and the detection of design errors before implementation. In the questionnaire, advanced simulations were measured with two questions:

- *q14. In the enterprise, advanced computer simulations of production processes are conducted to enhance their operations,*
- *q15. In the enterprise, digital models (digital twin) are built using advanced simulations for prototyping and process improvement.*

Autonomous robots are the new generation of industrial robots and artificial intelligence that can collaborate with humans through interaction, communicate with each other, and learn from historical data. Five questions were posed in the questionnaire:

- *q16. In the enterprise, automated and robotized production lines (technological processes and/or production cells) are operational,*
- *q17. In the enterprise, autonomous IT-computer systems are in place, controlling their devices and responding to commands from external control systems (cloud-based applications),*
- *q18. In the enterprise, transport (traffic) scenarios are developed with object and vehicle detection projects (using drones for object monitoring),*
- *q19. In the enterprise, industrial and service robots are utilized in collaboration with humans (operators),*
- *q20. In the enterprise, autonomous vehicles (AGVs) are used in internal transport or logistics.*

Universal integration involves connecting systems both within the enterprise and across the entire value chain, creating a comprehensive system encompassing all departments and functions, which, through vertical and horizontal integration, creates a complete picture of the organization as a complex system composed of various components. Two questions were posed on this topic in the questionnaire:

- *q21. In the enterprise, computer communication and process handling systems are integrated with external user systems to streamline communication and document flow,*
- *q22. In the enterprise, a flexible production environment allowing real-time production corrections is operational (autonomous robots, autonomous control systems).*

Augmented reality is an interactive version of the real environment enriched with digital visual elements, sounds, and other sensory stimuli using holographic technology, which adds multimedia information to the reality perceived by humans using mobile, visual, auditory, or manipulative devices. Two questions were designed on this topic in the questionnaire:

- *q23. In the enterprise, Virtual Reality (VR) techniques are used for employee training,*
- *q24. In the enterprise, Augmented Reality (AR) techniques are used for project and service work.*

Additive manufacturing, also known as 3D printing, is the process of creating three-dimensional objects. 3D printing is used for prototyping, manufacturing special components, spare parts, and highly customized product batches. Three questions were used to measure additive manufacturing:

- *q25. In the enterprise, 3D printing is used for designing solutions and products for customers (commercial product printing),*
- *q26. In the enterprise, 3D printing is used to ensure equipment continuity (printing spare parts for own use),*
- *q27. In the enterprise, 3D printing is used to print prototypes (models) to shorten product design time.*

Cybersecurity is the protection of factory production systems, computer systems, networks, programs, and data from attacks, unauthorized access, damage, or theft. Two questions were posed regarding cybersecurity in Polish manufacturing companies:

- *q28. In the enterprise, dedicated cybersecurity systems are employed,*
- *q29. In the enterprise, security measures such as passwords, encryption, and codes are implemented to enhance data security.*

The measurement of the 29 questions related to the nine pillars of Industry 4.0 was conducted using an ordinal scale. Measuring phenomena using an ordinal scale means arranging them according to the intensity of this feature. The ordinal scale allows establishing the relationships of 'more' or 'less' between objects that have a given feature to varying degrees, or 'equality' when they are identical in terms of this feature. Measurement is done by assessing the property using an ordered nomenclature. It is possible to arrange categories according to the degree to which they possess a certain feature. A popular symbol in ordinal measurement is numbers called ranks.

The construction of the ordinal scale for assessing the degree of business maturity of manufacturing enterprises in the context of Industry 4.0 was based on several assumptions. First, experiences in measuring business maturity on a scale from 1 to 5 used in typical maturity models (Alabbadi et al., 2023; Glaveli et al., 2023; Skyttermoen and Wedum, 2023; Maris et al., 2023; Kozlov et al., 2023) were utilized. Additionally, modified scale descriptions used to assess the maturity levels of quality management systems (PN-ISO 10014:2008) were used. Due to the multitude of innovative solutions related to Industry 4.0, the scale construction included level zero, indicating that a given practice does not occur at all. Ultimately, the scale has the following levels: 0; 1; 2; 3; 4; 5. The scale descriptions are as follows:

- **Rating 1:** The practice is established but has not yet been taken up, very little is happening (occurs in 1–20% of cases). Industry 4.0 solutions are rarely used, with some initial solutions from this scope possibly present;
- **Rating 2:** The practice is only noticeable in some areas (occurs in 21–40% of cases). Industry 4.0 solutions are used within the given pillar to a moderate extent. Their use is 'isolated', affecting only selected processes/products/technologies;
- **Rating 3:** The practice is widely established, but not in the majority of areas (occurs in 41–60% of cases). Industry 4.0 solutions are used moderately. Their application covers about half of the processes/products/technologies present in the organization;
- **Rating 4:** The practice is very typical, with only some exceptions (occurs in 61–80% of cases). Industry 4.0 solutions are used within the given pillar extensively. Their use is comprehensive, affecting most processes/products/technologies;
- **Rating 5:** The practice is implemented throughout the organization, virtually without exceptions (occurs in 81–100% of cases). The organization fully and comprehensively uses solutions from the given Industry 4.0 pillar across all activities in the entire value chain;
- **Rating 0:** The practice does not occur at all.

The questionnaire included a demographic section with questions about the company, including: industry, employment size, year of establishment, and type of capital (Table 6).

In scientific research, tools consisting of multiple variables are created to study specific phenomena. These variables can be indicators of varying intensity for the given phenomenon. Sometimes, certain variables are key to the scale, while others may reduce its cognitive value and should be removed during research and analysis. Variables in the scale must measure the same theoretical construct and should be as consistent as possible. This consistency is referred to as the reliability of the scale. The higher the reliability, the greater the chance of obtaining

**Table 6** List of variables used in the demographic section.

| *Variable Name* | *Range* | *Range of Variable Values* |
|---|---|---|
| Industry | Nominal | Automotive<br>Steel<br>Food |
| Employment size | Ordinal | 10–49 employees<br>50–249 employees<br>Above 250 employees |
| Year of establishment of the enterprise | Interval | Up to 1989<br>1990–2004<br>2005–2020 |
| Capital of the enterprise | Nominal | Polish<br>Foreign<br>Mixed |

*Source:* Own research.

the same results in subsequent measurements. Reliability analysis enables the examination of the properties of measurement scales and their components. One of the most commonly used measures of reliability is Cronbach's alpha, based on the correlation values between items (Cronbach, 1951). Cronbach's alpha coefficient relies on Pearson's R correlation. Through Cronbach's alpha index, questions in the scale that reduce its reliability can be identified. Cronbach's alpha values range from 0 to 1, with higher values indicating greater scale reliability. Typically, values above 0.7 are considered to indicate proper scale reliability.

To assess the reliability of the scale of variables defining the individual pillars of Industry 4.0, Cronbach's alpha coefficient was used. According to methodological principles in research, it was assumed that a set of indicators represents acceptably reliable measurement with Cronbach's alpha values > 0.7. The alpha coefficient value was determined based on the responses of the respondents participating in the research. The results of the reliability analysis are presented in Table 7.

**Table 7** Reliability statistics of variables in the questionnaire.

| *Pillars* | *N* | *Number of Items* | *Cronbach's Alpha* |
|---|---|---|---|
| Internet of Things | 317 | 4 | 0.910 |
| Big Data | 317 | 6 | 0.944 |
| Cloud computing | 317 | 3 | 0.917 |
| Advanced simulation | 317 | 2 | 0.918 |
| Autonomous systems | 317 | 5 | 0.914 |
| Universal integration | 317 | 2 | 0.807 |
| Virtual and Augmented Reality | 317 | 2 | 0.961 |
| Additive manufacturing/3D printing | 317 | 3 | 0.971 |
| Cybersecurity | 317 | 2 | 0.953 |

*Source:* Own research.

For all groups of variables constituting the individual pillars of Industry 4.0, the Cronbach's Alpha index was above 0.7. The highest index was recorded for the variables comprising the additive manufacturing pillar (0.971), and the lowest, but still acceptable, for the universal integration pillar (0.807). The reliability analysis indicates that there is no reason to exclude any of the variables. Additionally, a reliability analysis was conducted summarily for the entire tool consisting of 29 variables, and a Cronbach's Alpha value of 0.974 was obtained. This indicates very high scale reliability.

## 3. The Process of Direct Empirical Research

The research procedure undertaken focused on a sequence of 7 stages, as delineated in Fig. 2.

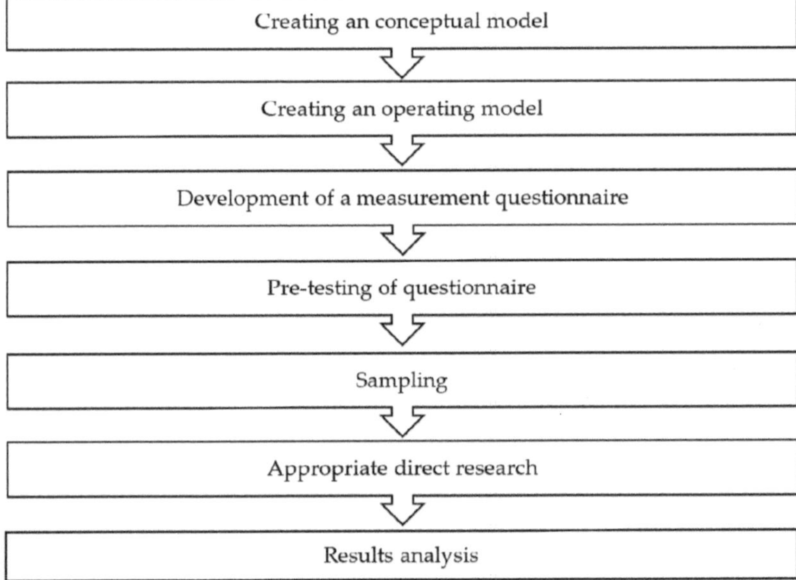


**Fig. 2**  The research procedure.
*Source:* Own study.

The research presented in the monograph was conducted using the methodology of an online survey. Conducting research using the internet poses a challenge for researchers, mainly due to the recruitment of respondents for online surveys. Electronic invitations to participate in online surveys are often perceived as spam. Therefore, the graphic presentation in the invitation plays a crucial role in attracting the attention of potential participants (Tomaselli et al., 2022). The effectiveness of the research and the reliability of the information obtained largely depend on the researchers' ability to control the distribution of the research sample. However, in survey research, full control over the characteristics of the sample remains a challenge despite efforts and reminders (Fang and Wen, 2012). Considering the issues related to conducting online research, the authors utilized the assistance of a professional research unit at the University of Economics in Katowice, the Center for Research and Development.

The prepared survey questionnaire was introduced to the online research platform SurveyMonkey.com. The initial questionnaire was subjected to preliminary research. Twelve enterprises were selected for the pre-test. The respondents rated the questionnaire according to the content and relevance of the items, and their opinions required minor corrections to improve the readability and understandability of the questionnaire. This pre-testing process allowed the development of a comprehensive and refined questionnaire.

Due to the subject of the research and the research technique, a purposive sampling method was chosen. Before starting the research, the minimum assumed research sample was determined. Calculations were performed based on the following formula:

$$n = \frac{N}{1 + \dfrac{4d^2(N-1)}{Z^2}}$$

where:
$N$ – population size,
$Z$ – standard value for the given confidence level,
$d$ – assumed estimation error.

The population size of manufacturing companies in Poland operating in the automotive, steel, and food industries was 18,908 in 2023. Considering an estimation error of 5% (0.05) and a confidence level of 90%, the minimum sample size obtained was 265 companies. Assuming that the research results for individual industries should be comparable, it was decided to distribute the sample as evenly as possible.

The research was conducted using the online survey technique, supported by telephone, from April 9 to May 6, 2024. Based on the national databases of manufacturing companies, 1,508 companies from across the country were invited to participate in the study. The research was conducted in Polish. The invitation page for the online survey is presented in Fig. 3.

The preamble used in the questionnaire, translated into English reads:

*Dear Sir or Madam,*

*We would like to invite you to participate in a scientific study on Industry 4.0 in manufacturing companies in Poland. The research is anonymous and the results in the form of aggregated studies will be used in a foreign scientific publication.*

*Thank you in advance for your time and responses.*

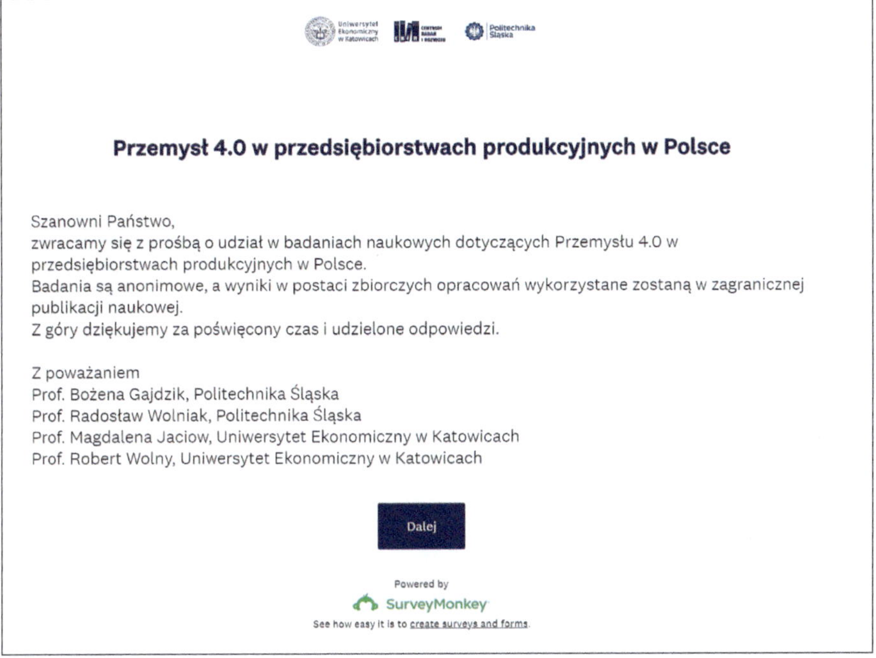



**Fig. 3**  Invitation to the research placed at the beginning of the questionnaire.
*Source:* Own study.

The research was supported by telephone, during which the purpose of the project, the thematic scope of the study, and encouragement to participate in the survey were presented during the conversations. The respondents were senior management personnel of the companies.

A total of 320 manufacturing companies from Poland participated in the research, of which 317 completed questionnaires were qualified for further work after formal analysis. The response rate was nearly 20%. Assuming a minimum sample size of 265 companies, the result of 317 was considered very satisfactory. The number of companies surveyed by industry was: 104 in the automotive industry, 103 in the steel industry, and 110 in the food industry.

As mentioned earlier, the research sample included almost an equal number of manufacturing companies from the surveyed industries. Almost 40% of the sample consisted of small companies, employing 10–49 employees, and nearly the same proportion of medium-sized companies, employing 50–249 employees. The remaining part of the sample consisted of large companies, employing 250 or more employees. Nearly half of the surveyed companies were established between 1990 and 2004, and almost one-third between 2005 and 2020. The oldest company in the sample was founded in 1880. The sample was dominated by companies with Polish capital (over 70%). Nearly one in four companies had mixed capital, and only 5.4% had foreign capital (Table 8).

**Table 8**   Sample of the industry enterprises (in %).

| *Items* | | % |
|---|---|---|
| Industry | Automotive | 32.8 |
| | Steel | 32.5 |
| | Food | 34.7 |
| Employment size | 10–49 employees | 39.1 |
| | 50–249 employees | 38.2 |
| | Above 250 employees | 22.7 |
| Year of establishment of the enterprise | Up to 1989 | 17.0 |
| | 1990–2004 | 45.5 |
| | 2005–2020 | 37.5 |
| Capital of the enterprise | Polish | 71.0 |
| | Foreign | 5.4 |
| | Mixed | 23.6 |

*Source:* Own research.

The collected empirical material was analyzed. All analyses of the data obtained were carried out using SPSS Statistics version 29.0, a specialized software for this purpose.

Measurement on the ordinal scale, which was used to assess the business maturity of Polish manufacturing companies, is carried out by measuring a property defined by ordered nomenclature. It is possible to arrange categories according to the degree to which they possess a certain feature. The arithmetic operation permissible on this scale is comparison. Measures of central tendency (structure indicators, mode, median, arithmetic mean) and cluster analysis were applied in data analysis.

The analysis of the structure of the studied phenomenon is a basic activity that should be performed before interpreting and drawing conclusions. The analysis of the structure of market phenomena involves identifying and interpreting the regularities existing in a given population. Structure indicators represent the share of parts of the statistical population in relation to the whole and are usually expressed as percentages. The mode (modal value, most frequent value) identifies the value of a variable that occurs most frequently in a given empirical distribution. For a series of counts and for a discrete variable, the mode is the value with the highest frequency. The median (middle quartile) divides the population into two parts such that 50% of the units have values lower and 50% higher than the middle quartile.

Cluster analysis is a data analysis tool used to group n objects, described by a vector of p features, into K non-empty, disjoint, and as 'homogeneous' as possible groups–clusters (Marek, 1989). The analysis is considered a tool for exploratory data analysis, the aim of which is to arrange objects into groups in such a way that the degree of association of objects within the same group is as high as possible, and with objects from other groups as low as possible (the principle of internal similarity and external dissimilarity). The results of cluster analysis are used to detect structures in the data. In the conducted analysis, hierarchical methods were used for classification. These methods are most often used in the clustering procedure. Clusters are determined by merging (agglomeration) smaller clusters formed in the previous steps of the algorithm or by splitting larger ones in divisive methods. To create a new cluster and determine its distance from other objects, Ward's method was used. In this method, distances between clusters are estimated using variance analysis (Wolny, 2016). The measure of distance between objects (clusters) is the within-group variance for the group formed by merging objects (clusters). A new cluster is formed at each stage from those objects (clusters) that ensure the minimization of within-group variance. In the literature, Ward's method is considered one of the most effective clustering methods (in creating homogeneous clusters), noting that it tends to create small-sized clusters. The agglomeration process is illustrated by a dendrogram, a binary tree whose nodes represent clusters and leaves represent individual objects.

The results of the analyses regarding the assessment of the business maturity of Polish manufacturing companies using measures of central tendency and cluster analysis will be presented in Chapter 4.

## References


Alabbadi, S., Seva-Larrosa, P. and García-Lillo, F. 2023. The mediating effect of information technology business strategic maturity on the relationship between organizational behavior and firm performance: A study of the Jordanian maritime industry. *Problems and Perspectives in Management, 21*(3): 753–763.

Cronbach, L.J. 1951. Coefficient alpha and the internal structure of tests. *Psychometrika, 16*: 297–334.

Fang, J. and Wen, C. 2012. Predicting potential respondents' decision to participate in web surveys. *Int. J. Serv. Technol. Manag., 18*: 16–32.

Glaveli, N., Alexiou, M., Maragos, A., Daskalopoulou, A. and Voulgari, V. 2023. Assessing the Maturity of Sustainable Business Model and Strategy Reporting under the CSRD Shadow. *Journal of Risk and Financial Management, 16*(10): 445.

Kozlov, A.V., Zaychenko, I.M. and Kolotova, D.P. 2023. Business Digital Maturity Assessment in Strategic Decision Making. *Lecture Notes in Networks and Systems, 684 LNNS*: 921–934.

Marek T. 1989. *Analiza skupień w badaniach empirycznych*. Metody SAHN, PWN, Warszawa.

Maris, A., Ongena, G. and Ravesteijn, P. 2023. Business Process Management Maturity and Process Performance: A Longitudinal Study. *Lecture Notes in Business Information Processing, 490 LNBIP*: 355–371.

PN-ISO 10014:2008. *Zarządzanie jakością. Wytyczne do osiągania korzyści finansowych i ekonomicznych*. PKN Warszawa.

Pollak, A. 2022. *APA Group, NAZCA 4.0. Przemysł 4.0 w praktyce*. https://open.spotify.com/show/4lJapGvA6aBm7QNbWIKoWe Dostęp: 24.11.2022SS), *IEEE*, 3928–3937.

Rüßmann, M., Lorenz, M., Gerbert, P., Waldner, M., Justus, J., Engel, P. and Harnisch, M. 2015. *Industry 4.0. The Future of Productivity and Growth in Manufacturing Industries*. The Boston Consulting Group (Issue April).

Skyttermoen, T. and Wedum, G. 2023. Developing Capabilities for Sustainable Business Models: Exploring Project Maturity for Innovation Processes. *Proceedings of the 19th European Conference on Management, Leadership, and Governance*, 2324 November, 370–379.

Tomaselli, V., Battiato, S., Ortis, A., Cantone, G.G., Urso, S. and Polosa, R. 2022. Methods, Developments, and Technological Innovations for Population Surveys. *Soc. Sci. Comput. Rev.*, *40*: 994–1013.

Wolny, R. 2016. The use of the cluster analysis to assess the functioning of cienams in Poland. *In*: M. Gajek, O. Hachkevych, A. Stanik-Besler, T. Wołczański (Eds.) *Manufacturing Proscesses. Actual Problems – 2016*. Volume I: *Basic Science Applications in Manufacturing Processes*, 155–168. University of Technology, Opole.

# 4 | Maturity Models of Manufacturing Companies
## Results of Own Research in Selected Polish Industries

This Chapter presents the results of direct research carried out on a sample of 317 Polish enterprises from the automotive, steel, and food industries. The distribution of assessments of the level of advancement of the implementation of Industry 4.0 technology was analyzed. The surveyed enterprises were categorized by industry. The obtained data were subjected to advanced statistical analysis. A cluster analysis was performed with the aim of identifying groups of technological factors within the nine pillars of Industry 4.0, which differ in the degree of implementation maturity in the surveyed enterprises. To conduct a cluster analysis of business maturity levels, one of the hierarchical grouping methods was used: the Ward's Method. In the Ward's Method, the distance between clusters is defined as the difference between the sums of the squares of deviations of individual units from the center of gravity of the groups to which these points belong. The distance between variables was determined based on the square of the Euclidean distance. The results presented in the chapter allow for a better understanding of which technologies are more or less advanced (mature) in the context of Industry 4.0 and how companies manage their implementation. The maturity models presented in the chapter are descriptive in nature.

## 1. Maturity Models in the Automotive Industry

The automotive industry consists of enterprises involved in the design, production, sale, and servicing of motor vehicles. Entities in the automotive sector include manufacturers of passenger cars, trucks, buses, and motorcycles; component manufacturers (engines, drive systems, braking systems, tires, batteries, and other parts); and companies providing services for vehicle users, such as auto repair shops, service centers, parts' stores, car dealerships, leasing companies, and Research and Development (R&D) units (conducting work on new technologies, innovations, and improvements in vehicle efficiency and safety).

Several key trends are driving the development of the automotive industry in the digital economy worldwide. One of these trends is the growing popularity of Electric Vehicles (EVs), driven by the goal of reducing $CO_2$ emissions and government support programs in many countries. At the same time, intensive research and development on autonomous vehicles could revolutionize transportation. The importance of sustainable production and recycling is also increasing, aiming to reduce environmental impact. The largest vehicle producers in the world are China, the USA, Japan, Germany, and South Korea. In terms of technological innovations, companies such as Tesla, Nissan, and BMW lead in electric vehicle production, and vehicles are increasingly equipped with communication technologies (IoT: Internet of Things), enabling better navigation and fleet management. Artificial Intelligence (AI) is used to develop autonomous driving systems and analyze data collected from vehicles.

In Poland, the automotive industry is also developing dynamically. Poland is one of the key players in the European automotive industry, with a large number of manufacturing plants for both cars and car parts, including factories owned by corporations such as Fiat, Volkswagen, Toyota, and Opel. Foreign investments in Poland are growing due to a competitive workforce and strategic location. Poland is developing infrastructure for charging electric vehicles, although the growth rate is slower than in some Western countries, and government initiatives and private investments are promoting the development of electromobility. Poland is becoming an important center for R&D in the automotive industry, attracting investments from global leaders, and numerous R&D centers are working on new technologies and innovations in the automotive sector. However, the automotive industry in Poland must meet challenges related to the transformation towards Industry 4.0 and the increasing demands for sustainable development, and key to the further growth of the sector will be infrastructure development and regulatory support.

The results of conducted studies allowed for the identification of the level of business maturity of automotive industry enterprises in the context of Industry 4.0. The studies included 100 enterprises from the automotive sector. Among this group, 33% were small enterprises employing 10–49 employees, 37% were medium-sized enterprises employing 50–249 employees, and 28% were large enterprises employing more than 250 people. Nearly 50% of them have been operating in the Polish market for 20 years, another 50%half even longer, and a few for over 30 years. Most of the surveyed companies are enterprises with Polish capital (Table 1).

The level of maturity of the automotive industry enterprises was identified in accordance with the conceptual and operational assumptions of the measurement model and the rating scale described in Chapter 3. Table 2 presents the percentage distribution of declarations of representatives of companies from the surveyed industry regarding the level of implementation of the Internet of Things (IoT) (pillar 1 – P1) on a scale of 0–5, where: 1 - the practice is established but has not yet been taken up, very little is happening (occurs in 1–20% of cases), 2 - the

**Table 1**  Sample of the automotive industry enterprises (in %).

| *Items* | | % |
|---|---|---|
| Employment size | 10–49 employees | 33.7 |
| | 50–249 employees | 37.5 |
| | Above 250 employees | 28.8 |
| Year of establishment of the enterprise | Up to 1989 | 12.5 |
| | 1990–2004 | 39.4 |
| | 2005–2017 | 48.1 |
| Capital of the enterprise | Polish | 63.5 |
| | Foreign | 10.5 |
| | Mixed | 26.0 |

*Source:* Own research.

practice is only noticeable in some areas (occurs in 21–40% of cases), 3 - the practice is widely established, but not in the majority of areas (occurs in 41–60% of cases), 4 - the practice is very typical, with only some exceptions (occurs in 61–80% of cases), 5 - the practice is implemented throughout the organization, virtually without exceptions (occurs in 81–100% of cases), and 0 - the practice does not occur at all.

The data indicates a high level of implementation of intelligent communication and visualization systems. A significant portion of automotive enterprises use intelligent communication systems, visualization systems, and process response systems, such as Andon systems. Over 82.7% of companies rate these practices at level 3 or higher, indicating that they are widely implemented and used in most areas of operation. There is also an advanced implementation of technical-business IoT solutions. Technical-business solutions based on IoT capabilities and/or industrial platforms are well integrated into enterprises. Nearly 83.7% of companies rate their implementation at level 3 or higher, demonstrating their significant role in business and production activities. RFID technology also plays a significant role. RFID technology and product identification codes are widely used for supply chain, production, logistics, and warehousing control. Over 75% of enterprises rate these practices at level 3 or higher, highlighting their importance in operational management. There is also a significant implementation of machine-to-machine (M2M) communication. Over 74.9% of companies rate these practices at level 3 or higher, indicating their crucial role in automation and operational efficiency.

The average level of implementation of pillar P1 in the surveyed enterprises reaches 40–70% of the activities undertaken in production processes (Table 3).

The level of advancement of the implementation of the Big Data (P2) pillar in the surveyed automotive companies is presented in Table 4. The data shows that a significant proportion of companies track machine parameters and vehicle locations

**Table 2**  Assessment of the level of internet of things implementation in the automotive industry enterprises (in %).

| Internet of Things | Rating | | | | | |
|---|---|---|---|---|---|---|
| | *0* | *1* | *2* | *3* | *4* | *5* |
| q1. In the enterprise, intelligent communication, visualization, and process response systems are operational (e.g., Andon system/boards) | 2.9 | 8.7 | 5.8 | 35.6 | 30.8 | 16.3 |
| q2. In the enterprise, technical-business solutions based on the capabilities of the Internet of Things (IoT) and/or industrial platforms have been implemented | 1.9 | 6.7 | 7.7 | 46.2 | 19.2 | 18.3 |
| q3. In the enterprise, Radio Frequency Identification (RFID) and product identification codes are used to control supply chain/production/logistics/warehouse | 1.9 | 8.7 | 13.5 | 25.0 | 30.8 | 20.2 |
| q4. In the enterprise, direct Machine-to-Machine (M2M) communication is facilitated using distributed device sensors and other components of industrial networks | 1.9 | 6.7 | 16.3 | 39.4 | 24.0 | 11.5 |

*Source:* Own research.

**Table 3**  Assessment of the level of P1 implementation in the automotive industry enterprises – selected statistics.

| $P1$ | $\bar{x}$ | $M_e$ | $M_o$ |
|---|---|---|---|
| $q1$ | 3.32 | 3 | 3 |
| $q2$ | 3.29 | 3 | 3 |
| $q3$ | 3.35 | 4 | 4 |
| $q4$ | 3.11 | 3 | 3 |

*Source:* Own research.

to improve processes, as confirmed by 41.3% of companies assessing this aspect at level 3, and 24% at level 4. Software for real-time data processing and analysis is also applicable. One in three companies rates this aspect at level 3, and the fourth at level 4. This is considered a significant technological advancement in this area. Dashboards are also widely used in the surveyed companies to monitor processes and make decisions. Every third company assesses the level of implementation of this Big Data technology at level 3, and the same number at level 4. It can be considered that this practice is well established in Polish enterprises. Most companies effectively use data to manage processes at scale by monitoring KPIs. Such monitoring is implemented in 37.5% at level 3 and in 33.7% at level 4. The level of implementation of Big Data algorithms in the diagnostics of production processes and self-diagnostics of machines is similar. Every third company has implemented this technology at level 3, the same number at level 4. Analysis of large datasets to determine customers' shopping preferences is also common in the surveyed companies in the automotive industry. More than 40% of companies rate the level of technology adoption as level 3, and 26% as level 4.

The average level of implementation of the P2 pillar in the surveyed companies reaches 40–70% of activities undertaken in production processes (Table 5).

The declared level of cloud technology implementation in the automotive industry enterprises is summarized in Table 6. The survey results indicate that companies are eager to use cloud technologies. Nearly two-thirds of the surveyed enterprises declare an implementation level of at least 60% for this technology. Companies are also keen to use data-based services that are available on cloud platforms. As many as 37.5% of companies rate this aspect at level 3, 26.9% at level 4, and 23.1% at level 5. This indicates a high level of implementation of these technologies, allowing for the efficient use of data in business operations. Most of the surveyed companies effectively utilize the flexibility and scalability offered by cloud technologies. Over 80% of the surveyed enterprises rate the use of system scalability technologies, supported by cloud data, at a level of at least 3.

**Table 4** Assessment of the level of Big Data implementation in the automotive industry enterprises (in %).

| Big Data | Rating | | | | | |
|---|---|---|---|---|---|---|
| | 0 | 1 | 2 | 3 | 4 | 5 |
| q5. In the enterprise, machine parameters and vehicle locations are tracked to streamline processes | 1.0 | 3.8 | 12.5 | 41.3 | 24.0 | 17.3 |
| q6. In the enterprise, software (computer system) is utilized for real-time processing and analysis of data on processes/machines/products | – | 5.8 | 13.5 | 36.5 | 27.9 | 16.3 |
| q7. In the enterprise, managerial dashboards (providing access to current information at the managerial level) are used to processes control (monitoring) and make decisions | 1.9 | 4.8 | 12.5 | 32.7 | 32.7 | 15.4 |
| q8. In the enterprise, data on key performance indicators (KPIs) monitoring process parameters are collected on a large scale (management system) | 1.9 | 7.7 | 4.8 | 37.5 | 33.7 | 14.4 |
| q9. In the enterprise, Big Data algorithms are employed for diagnosing production processes and machine self-diagnosis (maintenance – PM) | 5.8 | 5.8 | 11.5 | 36.5 | 32.7 | 7.7 |
| q10. In the enterprise, large datasets are analyzed to determine customer purchasing preferences and customer profile | 5.8 | 6.7 | 9.6 | 41.3 | 26.0 | 10.6 |

**Table 5**  Assessment of the level of P2 implementation in the automotive industry enterprises – selected statistics.

| P2 | $\bar{x}$ | $M_e$ | $M_o$ |
|---|---|---|---|
| *q5* | 3.36 | 3 | 3 |
| *q6* | 3.36 | 3 | 3 |
| *q7* | 3.36 | 3 | 3 |
| *q8* | 3.36 | 3 | 3 |
| *q9* | 3.08 | 3 | 3 |
| *q10* | 3.07 | 3 | 3 |

*Source:* Own research.

**Table 6**  Assessment of the level of cloud computing implementation in the automotive industry enterprises (in %).

| *Cloud Computing* | *Rating* | | | | | |
|---|---|---|---|---|---|---|
| | *0* | *1* | *2* | *3* | *4* | *5* |
| q11. In the enterprise, data is collected and stored in the cloud | 1.0 | 1.0 | 6.7 | 29.8 | 37.5 | 24.0 |
| q12. In the enterprise, access to various data-based services is provided, and these data are made available on cloud platforms | 1.0 | 2.9 | 8.7 | 37.5 | 26.9 | 23.1 |
| q13. In the enterprise, data collected and processed in the cloud are used for easy scaling of process support computer systems | 1.0 | 3.8 | 13.5 | 31.7 | 28.8 | 21.2 |

*Source:* Own research.

The average level of implementation of the P3 pillar in the surveyed companies reaches the level of 60–80% of activities undertaken in production processes (Table 7).

Based on the data presented in Table 8 regarding the level of implementation of advanced simulations in automotive industry enterprises, it can be concluded that advanced computer simulations of production processes are widely used in automotive companies. As many as 50% of companies rate this aspect at level 3, indicating that this practice is widespread, though not in all areas. Additionally, 22.1% of companies rate it at level 4, and 7.7% at level 5, which signifies a high level of advancement and substantial adaptation of these technologies to improve production operations. The use of advanced simulations to create digital twins for prototyping and process improvement is also well implemented. 48.1% of

**Table 7** Assessment of the level of P3 implementation in the automotive industry enterprises – selected statistics.

| P3 | $\bar{x}$ | $M_e$ | $M_o$ |
|---|---|---|---|
| q11 | 3.74 | 4 | 4 |
| q12 | 3.56 | 3 | 3 |
| q13 | 3.47 | 3 | 3 |

*Source:* Own research.

companies rate the level of implementation of this technology at level 3, and 14.4% at level 4, suggesting that nearly 50% of automotive enterprises use this technology extensively.

**Table 8** Assessment of the level of advanced simulation implementation in the automotive industry enterprises (in %).

| Advanced Simulation | Rating | | | | | |
|---|---|---|---|---|---|---|
| | 0 | 1 | 2 | 3 | 4 | 5 |
| q14. In the enterprise, advanced computer simulations of production processes are conducted to enhance their operations | 1.9 | 9.6 | 8.7 | 50.0 | 22.1 | 7.7 |
| q15. In the enterprise, digital models (digital twin) are built using advanced simulations for prototyping and process improvement | 5.8 | 10.6 | 13.5 | 48.1 | 14.4 | 7.7 |

*Source:* Own research.

The average level of implementation of the P4 pillar in the surveyed companies reaches the level of 30–50% of activities undertaken in production processes (Table 9).

**Table 9** Assessment of the level of P4 implementation in the automotive industry enterprises – selected statistics.

| P4 | $\bar{x}$ | $M_e$ | $M_o$ |
|---|---|---|---|
| q14 | 3.04 | 3 | 3 |
| q15 | 2.78 | 3 | 3 |

*Source:* Own research.

The declared level of implementation of autonomous systems in automotive industry companies is presented in Table 10. It shows that automated and robotic production lines are widely used in the automotive industry. One-third of companies have implemented this technology at level 3, one-quarter rate the implementation at level 4, and one-tenth at level 5. Autonomous IT systems that control devices and respond to commands from external control systems (cloud-based) are also implemented at a similar level. Fewer enterprises have implemented transport scenarios developed using object and vehicle detection technologies, including drones. One-third of enterprises rate the implementation of this technology at level 3. At the same time, one-quarter enterprises have not implemented this technology at all. Industrial and service robots, collaborating with humans, are quite commonly used in automotive industry enterprises. 28.8% of companies rate this aspect at level 3, 24% at level 4, and 7.7% at level 5. Autonomous Guided Vehicles (AGVs) used in internal transport or logistics are less commonly implemented. One-third of the surveyed automotive enterprises do not yet use these technologies (Table 10).

**Table 10**  Assessment of the level of autonomous systems implementation in the automotive industry enterprises (in %).

| *Autonomous Systems* | *Rating* | | | | | |
|---|---|---|---|---|---|---|
| | *0* | *1* | *2* | *3* | *4* | *5* |
| q16. In the enterprise, automated and robotized production lines (technological processes and/or production cells) are operational | 1.9 | 10.6 | 14.4 | 36.5 | 25.0 | 11.5 |
| q17. In the enterprise, autonomous IT-computer systems are in place, controlling their devices and responding to commands from external control systems (cloud-based applications) | 5.8 | 9.6 | 13.5 | 32.7 | 26.9 | 11.5 |
| q18. In the enterprise, transport (traffic) scenarios are developed with object and vehicle detection projects (using drones for object monitoring) | 26.9 | 7.7 | 10.6 | 37.5 | 11.5 | 5.8 |
| q19. In the enterprise, industrial and service robots are utilized in collaboration with humans (operators) | 6.7 | 6.7 | 26.0 | 28.8 | 24.0 | 7.7 |
| q20. In the enterprise, autonomous vehicles (AGVs) are used in internal transport or logistics | 34.6 | 8.7 | 6.7 | 33.7 | 10.6 | 5.8 |

*Source:* Own research.

The average level of implementation of pillar P5 in the surveyed enterprises reaches 20–40% of the activities undertaken in production processes (Table 11).

**Table 11**  Assessment of the level of P5 implementation in the automotive industry enterprises – selected statistics.

| P5 | $\bar{x}$ | $M_e$ | $M_o$ |
|---|---|---|---|
| q16 | 3.07 | 3 | 3 |
| q17 | 3.00 | 3 | 3 |
| q18 | 2.16 | 3 | 3 |
| q19 | 2.79 | 3 | 3 |
| q20 | 1.94 | 2 | 0 |

*Source:* Own research.

Table 12 contains data on the declared level of implementation of universal integration technologies in automotive industry enterprises. The data shows that the integration of computer communication and process management systems with external user systems is widely used in automotive companies. 27.9% of firms rate the implementation of this technology in their company at level 3, slightly more (32.7%) rate it at level 4, and 23.1% at level 5. A flexible production environment, which allows for real-time production adjustments (including autonomous robots and control systems), is also well implemented. 41.3% of companies rate this aspect at level 3, 24% of companies rate it at level 4, and 7.7% at level 5.

**Table 12**  Assessment of the level of universal integration implementation in the automotive industry enterprises (in %).

| Universal Integration | Rating | | | | | |
|---|---|---|---|---|---|---|
| | 0 | 1 | 2 | 3 | 4 | 5 |
| q21. In the enterprise, computer communication and process handling systems are integrated with external user systems to streamline communication and document flow | 1.0 | 7.7 | 7.7 | 27.9 | 32.7 | 23.1 |
| q22. In the enterprise, a flexible production environment allowing real-time production corrections is operational (autonomous robots, autonomous control systems) | 1.9 | 11.5 | 13.5 | 41.3 | 24.0 | 7.7 |

*Source:* Own research.

The average level of implementation of pillar P6 in the surveyed enterprises reaches 40–50% of the activities undertaken in production processes (Table 13).

**Table 13** Assessment of the level of P6 implementation in the automotive industry enterprises – selected statistics.

| P6 | $\bar{x}$ | $M_e$ | $M_o$ |
|---|---|---|---|
| q21 | 3.53 | 4 | 4 |
| q22 | 2.97 | 3 | 3 |

*Source:* Own research.

The utilization of Virtual Reality (VR) and Augmented Reality (AR) within automotive industry enterprises remains limited. The data presented in Table 14 indicates a moderate level of VR technology implementation in employee training. Only two-fifths out of the surveyed enterprises evaluate this aspect of technology implementation at level 3, with nearly one-sixth rating it at level 4. Conversely, AR techniques are employed in design and service work by one-third of enterprises at level 3, while one-fifth of companies rate their implementation at level 4. Furthermore, it is noteworthy that one-fifth automotive enterprises do not utilize any of these technologies (VR or AR).

**Table 14** Assessment of the level of Virtual and Augmented Reality implementation in the automotive industry enterprises (in %).

| *Virtual and Augmented Reality* | *Rating* | | | | | |
|---|---|---|---|---|---|---|
| | *0* | *1* | *2* | *3* | *4* | *5* |
| q23. In the enterprise, Virtual Reality (VR) techniques are used for employee training | 19.2 | 12.5 | 11.5 | 39.4 | 17.3 | – |
| q23. In the enterprise, Augmented Reality (AR) techniques are used for project and service work | 19.2 | 11.5 | 13.5 | 36.5 | 19.2 | – |

*Source:* Own research.

The average level of implementation of the P7 pillar in the surveyed companies reaches 20–40% of activities undertaken in production processes (Table 15).

**Table 15** Assessment of the level of P7 implementation in the automotive industry enterprises – selected statistics.

| P7 | $\bar{x}$ | $M_e$ | $M_o$ |
|---|---|---|---|
| *q23* | 2.23 | 3 | 3 |
| *q24* | 2.25 | 3 | 3 |

*Source:* Own research.

The level of implementation of 3D printing technology within automotive industry enterprises is not high (Table 16). Nearly one-fifth enterprises do not utilize this technology. 3D printing for designing solutions and products for customers is employed by 36.5% of enterprises at level 3, and by 22.1% of firms at level 4. Conversely, 3D printing to ensure equipment continuity (printing spare parts) is implemented at level 3 by 45.2% of enterprises, and at level 4 by 11.5%. Additionally, 3D printing is utilized for creating prototypes (models) to reduce design time. This process is conducted at level 3 by one-third of the surveyed enterprises, and at level 4 by 17.3%.

**Table 16**  Assessment of the level of additive manufacturing/3D printing implementation in the automotive industry enterprises (in %).

| *Additive Manufacturing/3D Printing* | *Rating* | | | | | |
|---|---|---|---|---|---|---|
| | *0* | *1* | *2* | *3* | *4* | *5* |
| q25. In the enterprise, 3D printing is used for designing solutions and products for customers (commercial product printing) | 18.3 | 6.7 | 14.4 | 36.5 | 22.1 | 1.9 |
| q29. In the enterprise, 3D printing is used to ensure equipment continuity (printing spare parts for own use) | 18.3 | 7.7 | 16.3 | 45.2 | 11.5 | 1.0 |
| q27. In the enterprise, 3D printing is used to print prototypes (models) to shorten product design time | 20.2 | 7.7 | 18.3 | 35.6 | 17.3 | 1.0 |

*Source:* Own research.

The average level of implementation of the P8 pillar in the surveyed companies reaches 20–40% of activities undertaken in production processes (Table 17).

**Table 17**  Assessment of the level of P8 implementation in the automotive industry enterprises – selected statistics.

| *P8* | $\bar{x}$ | $M_e$ | $M_o$ |
|---|---|---|---|
| *q25* | 2.43 | 3 | 3 |
| *q26* | 2.69 | 3 | 3 |
| *q27* | 2.50 | 3 | 3 |

*Source:* Own research.

Automotive industry enterprises report a high level of implementation of cybersecurity technologies, including both dedicated cybersecurity systems and basic data protection measures such as passwords, encryption, and codes

(Table 18). Dedicated cybersecurity systems are widely utilized in automotive enterprises, with 30.8% of companies rating their implementation at level 3, and over 50% rating it at a minimum of level 5. Only 3.8% of companies have not yet implemented such systems. A similar level of implementation applies to available security measures, such as passwords, encryption, and codes. 36.5% of enterprises rate this aspect at level 3, 8.7% at level 4, and 44.2% at level 5.

The average level of implementation of the P9 pillar in the surveyed companies reaches the level of 60–80% of activities undertaken in production processes (Table 19).

**Table 18**  Assessment of the level of cybersecurity implementation in the automotive industry enterprises (in %).

| Cybersecurity | Rating | | | | | |
|---|---|---|---|---|---|---|
| | *0* | *1* | *2* | *3* | *4* | *5* |
| q28. In the enterprise, dedicated cybersecurity systems are employed | 3.8 | 3.8 | 3.8 | 30.8 | 17.3 | 40.4 |
| q29. In the enterprise, security measures such as passwords, encryption, and codes are implemented to enhance data security | 1.0 | 3.8 | 5.8 | 36.5 | 8.7 | 44.2 |

*Source:* Own research.

**Table 19**  Assessment of the level of P9 implementation in the automotive industry enterprises – selected statistics.

| P9 | $\bar{x}$ | $M_e$ | $M_o$ |
|---|---|---|---|
| q28 | 3.75 | 4 | 5 |
| q29 | 3.81 | 4 | 5 |

*Source:* Own research.

In summary, the compiled data indicates that automotive industry enterprises in Poland exhibit a relatively high level of advancement in implementing Industry 4.0 technologies. This is particularly evident in key areas such as the Internet of Things (IoT), Big Data, cloud technologies, advanced simulations, and cybersecurity systems. A high percentage of firms implementing these practices at a minimum level of 40%, demonstrates a broad adoption of modern technologies, significantly contributing to improved efficiency, management, and automation of processes within the industry. Despite the advanced implementation of certain technologies, such as IT systems and robotic production lines, there are still areas with substantial potential for further development, including autonomous AGVs, transport scenarios utilizing drones, and VR and AR technologies. Additionally, the well-implemented integration of computer communication systems and flexible production environments reflects the industry's drive to enhance operational efficiency through advanced technological tools. Overall, automotive enterprises in Poland show significant progress in integrating Industry 4.0 technologies, although challenges and opportunities for further investment and adaptation remain.

The highest level of business maturity in the context of implementing Industry 4.0 technologies in automotive enterprises is observed in pillars 9 and 3, while the lowest is in pillars 7 and 8 (Figure 1). A higher level of maturity is reported by automotive enterprises with foreign capital compared to those with Polish or mixed capital. The declared level of digital technology implementation is also higher in enterprises that have been in the market since at least 2005, compared to those with a longer market presence. A higher level of digital maturity is also observed in the largest enterprises (employing over 250 employees), with slightly lower levels in small and medium-sized enterprises. It is characteristic of the automotive industry that, regardless of the features of the surveyed enterprises (size, capital origin, length of market presence), the highest level of technology implementation pertains to pillar 9, which is cybersecurity. Nearly all surveyed companies recognize the importance of this pillar and strive to implement cybersecurity technologies at the highest possible level.

The objective of the cluster analysis was to identify groups of technological factors within the nine pillars of Industry 4.0 that differ in their degree of implementation maturity in automotive industry enterprises. This allows for a better understanding of which technologies are more or less advanced (mature) in the context of Industry 4.0, and how enterprises are managing their implementation.

To conduct the cluster analysis of the business maturity levels of manufacturing enterprises in the automotive industry in the context of Industry 4.0, one of the

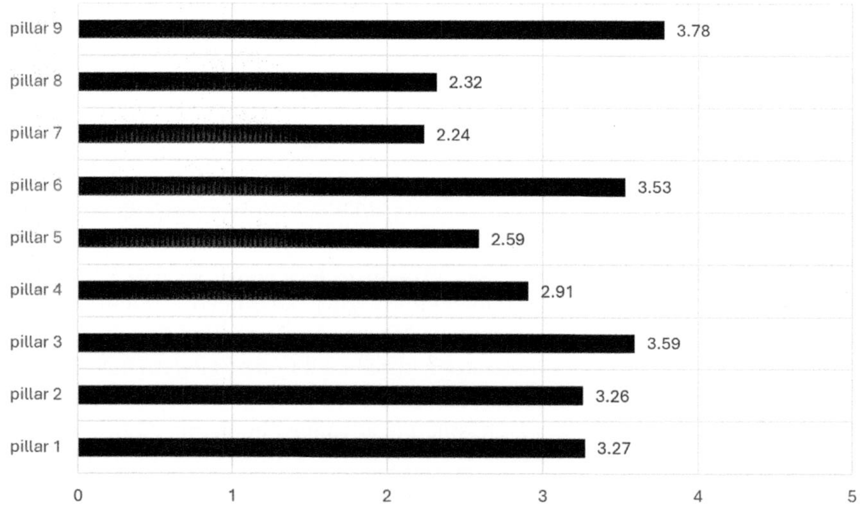


**Fig. 1** Assessment of the level of maturity of pillars of industry 4.0 in the automotive industry enterprises – mean values.
*Source:* Own research.

hierarchical clustering methods—the Ward's method—was applied. In Ward's method, the distance between clusters is defined as the difference between the sums of squared deviations of individual units from the centroid of the groups to which these points belong. The distance between variables was determined based on the squared Euclidean distance.

Based on the analysis conducted for the assessments of the various pillars of Industry 4.0, three clusters were obtained (Figure 2). Within each of these three categories, the characteristics of the individual pillars of Industry 4.0 are similar in the context of the maturity assessment by the respondents, whereas between

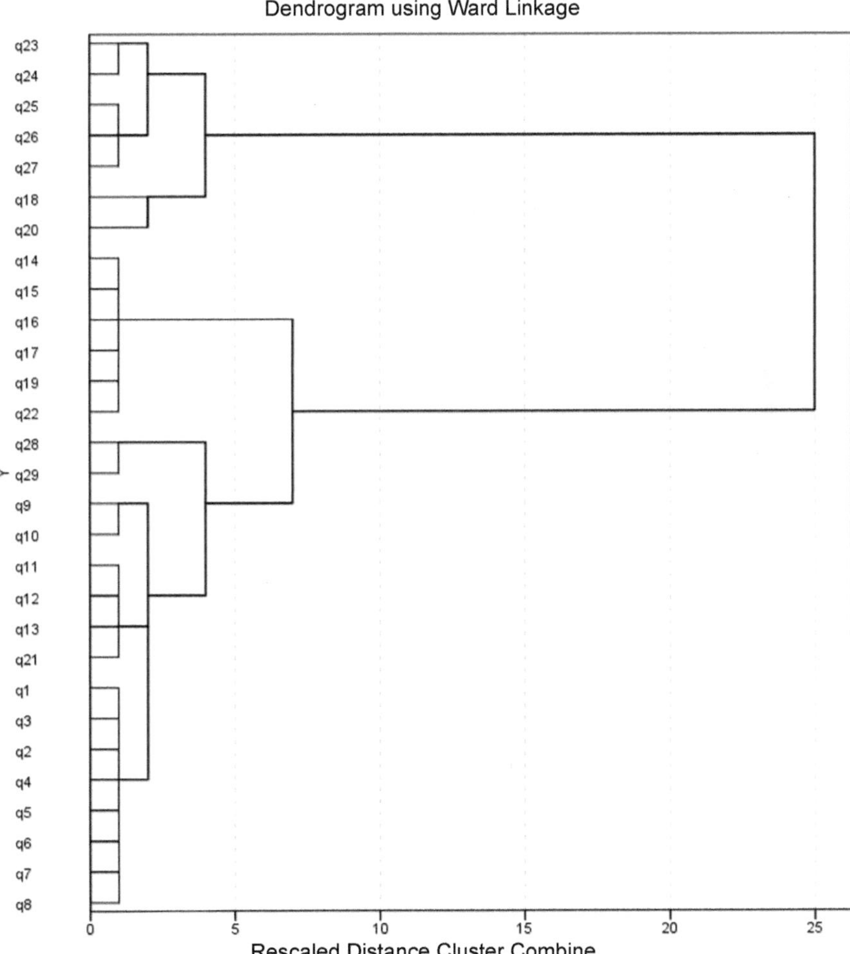



**Fig. 2** Classification of business maturity characteristics in manufacturing enterprises in the automotive industry in the context of industry 4.0.
*Source:* Own research.

the categories they differentiate from each other. For instance, entrepreneurs evaluating the maturity level related to VR and AR similarly rated the maturity level related to additive manufacturing and autonomous systems, while they rated the maturity level differently for advanced simulations and cybersecurity.

Cluster 1 includes seven factors characterized by the lowest average scores of business maturity. These factors are related to advanced technologies such as VR and AR (q23, q24), selected autonomous systems (q18, q20), and additive manufacturing/3D printing (q25, q26, q27). These are areas where automotive industry enterprises are least mature and are even considered to face the greatest difficulties in implementing these Industry 4.0 technologies. Within this cluster, the technologies in the area of autonomous systems, specifically transport scenarios that involve projects using drones for object monitoring and the use of autonomous vehicles in internal transport, are at the lowest level of maturity. The low variability of scores in this cluster indicates uniformity in the low level of technological advancement (Figure 3).

Cluster 2, encompassing six factors, presents an intermediate level of business advancement. The factors in this cluster are primarily technologies related to advanced simulation (q14, q15), selected autonomous systems (q16, q17, q19), and a flexible production environment that allows real-time production adjustments (q22). Automotive industry enterprises have similarly rated the implementation status of these technologies at a moderate level, which may indicate that they are undergoing digital transformation in this area. The most closely rated factors in terms of maturity pertain to autonomous IT-computer systems that control their own devices and respond to commands from external control systems, as

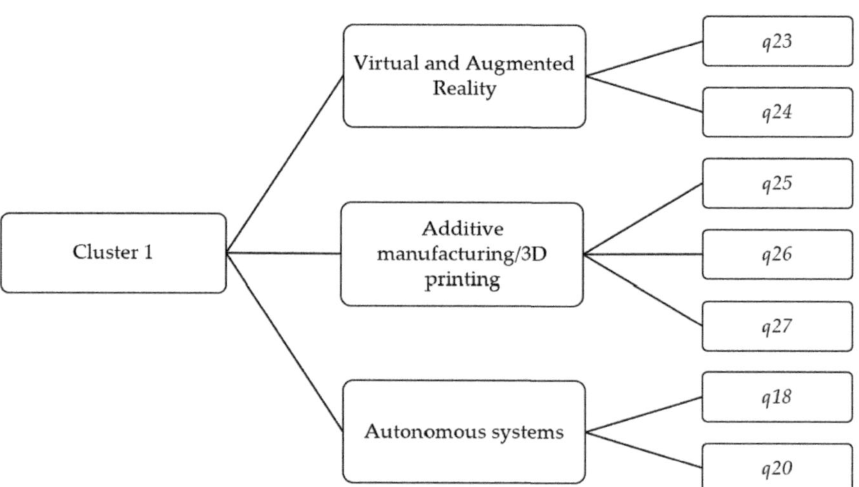


**Fig. 3** Variables identified within cluster 1 of manufacturing enterprises in the automotive industry.
*Source:* Own research.

well as the flexible production environment that enables real-time production adjustments. There is a significant need for further development and support in the implementation of these technologies, requiring investments in technological infrastructure and the development of employee competencies in areas such as operational handling. The average variability of ratings in this cluster suggests some variation in the level of advancement, but it still indicates a relatively homogeneous group in terms of technological maturity (Figure 4).

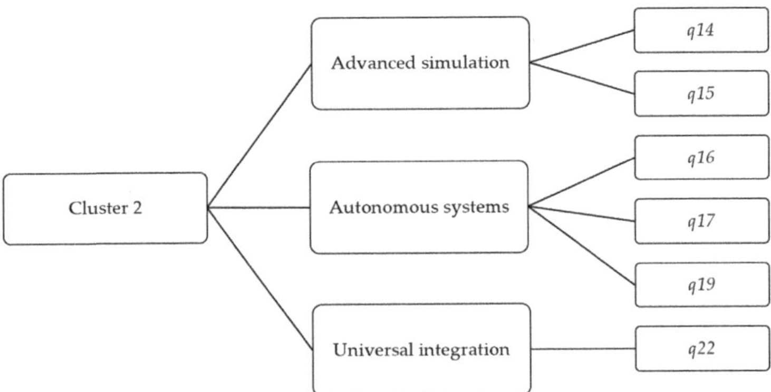


**Fig. 4** Variables identified within cluster 2 of manufacturing enterprises in the automotive industry.
*Source:* Own research.

Cluster 3, encompassing 16 factors, is characterized by the highest average scores of business maturity in the context of Industry 4.0. The dominant factors in this cluster include advanced technologies such as Big Data (q5–q10), IoT (q1–q4), cloud computing (q11–q13), one factor related to universal integration (q21), and the level of security related to cybersecurity (q28, q29). It can be considered that in these areas, automotive industry enterprises excel in the maturity of their implementation and are leaders in the deployment of these technologies. The highest maturity within Cluster 3, which is similarly rated, is observed in technologies related to cybersecurity and those associated with data collection and storage in the cloud. The high variability of scores in this cluster suggests significant diversity in the level of advancement of individual factors, which may result from varied approaches to technology implementation across different enterprises (Figure 5).

Cluster analysis provides valuable insights into the distribution of technological maturity within the automotive industry, enabling a better understanding of where to direct supportive actions and investments to accelerate digital transformation within the framework of Industry 4.0.

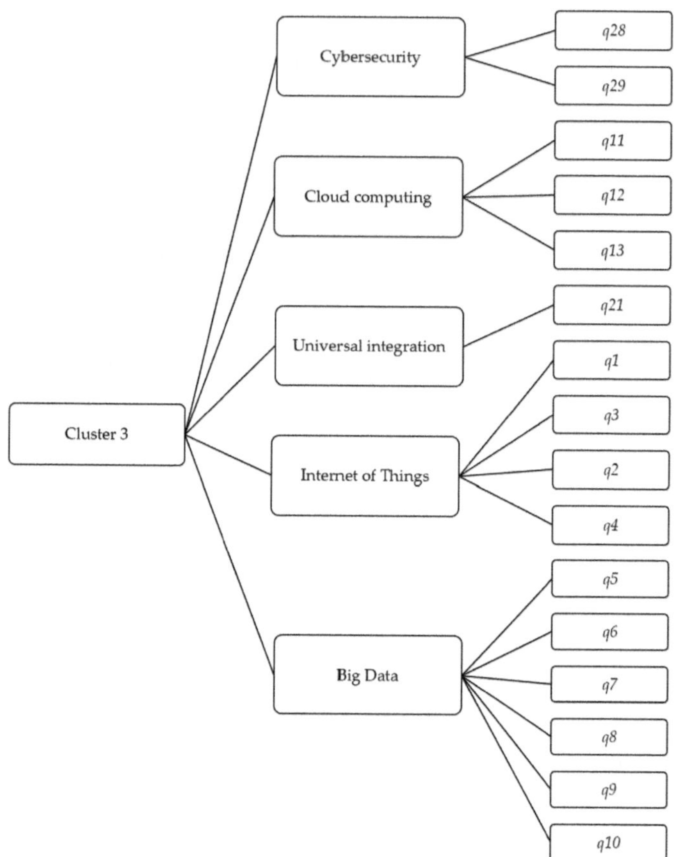


**Fig. 5**   Variables identified within cluster 3 of manufacturing enterprises in the automotive industry.
*Source:* Own research.

## 2.   Maturity Models in the Steel Industry

The steel industry encompasses all activities related to the production, processing, and distribution of steel. The global steel market is one of the key sectors of heavy industry. The largest steel producers in the world are China, India, Japan, the USA, and Russia. China dominates the global market, producing over 50% of the world's steel. China is also the largest consumer of steel, driven by its rapidly developing construction and infrastructure sectors. Other countries with high steel consumption include India, the USA, Japan, and Germany.

The steel industry continually invests in technological innovations to increase production efficiency and reduce its environmental impact. Technologies such as steel recycling, the use of renewable energy, and reducing $CO_2$ emissions are key trends in the market. Poland, as one of the key steel producers in Europe, has a well-developed steel industry. The most important steel production plants are located in Katowice, Kraków, Częstochowa, and Dąbrowa Górnicza. The Polish steel market primarily serves the construction, automotive, and machinery industries. In recent years, there has been an increase in demand for steel due to infrastructure development. Polish steel companies are investing in modern technologies to increase their competitiveness in the international market. They are implementing ecological solutions and technologies for the automation of production processes.

The steel industry in Poland faces challenges such as changing environmental regulations, fluctuations in raw material prices, and competition from Asian producers. The steel industry is a key sector of the economy both globally and domestically. The development of the steel industry is characterized by dynamic changes driven by technological innovations and the growing demand for steel.

The results of conducted studies allowed for the identification of the level of business maturity of steel industry enterprises in the context of Industry 4.0. The studies included 100 enterprises from the steel industry. Among this group, 52.4% were small enterprises employing 10–49 employees, 35.9% were medium-sized enterprises employing 50–249 employees, and 11.7% were large enterprises employing over 250 people. Nearly one-third of them have been operating in the Polish market for 20 years, half have been operating even longer, and a few for over 30 years. The vast majority of the surveyed companies are enterprises with Polish capital (Table 20).

**Table 20**    Sample of the steel industry enterprises (in %).

| *Items* | | *%* |
|---|---|---|
| Employment size | 10–49 employees | 52.4 |
| | 50–249 employees | 35.9 |
| | Above 250 employees | 11.7 |
| Year of establishment of the enterprise | Up to 1989 | 18.4 |
| | 1990–2004 | 50.5 |
| | 2005–2018 | 31.1 |
| Capital of the enterprise | Polish | 80.6 |
| | Foreign | 1.9 |
| | Mixed | 17.5 |

*Source:* Own research.

The maturity level of enterprises in the steel industry was identified according to the conceptual and operational assumptions of the measurement model and rating scale described in Chapter 3. Table 21 presents the percentage distribution of declarations by representatives of the surveyed steel industry enterprises regarding the level of implementation of the Internet of Things (IoT) (pillar 1 – P1) on a scale

**Table 21**  Assessment of the level of internet of things implementation in the steel industry enterprises (in %).

| Internet of Things | Rating | | | | | |
|---|---|---|---|---|---|---|
| | 0 | 1 | 2 | 3 | 4 | 5 |
| q1. In the enterprise, intelligent communication, visualization, and process response systems are operational (e.g., Andon system/boards) | 26.2 | 13.6 | 7.8 | 10.7 | 22.3 | 19.4 |
| q2. In the enterprise, technical-business solutions based on the capabilities of the Internet of Things (IoT) and/or industrial platforms have been implemented | 25.2 | 10.7 | 16.5 | 16.5 | 16.5 | 14.6 |
| q3. In the enterprise, Radio Frequency Identification (RFID) and product identification codes are used to control supply chain/production/logistics/warehouse | 26.2 | 6.8 | 10.7 | 15.5 | 22.3 | 18.4 |
| q4. In the enterprise, direct Machine-to-Machine (M2M) communication is facilitated using distributed device sensors and other components of industrial networks | 31.1 | 6.8 | 6.8 | 20.4 | 18.4 | 16.5 |

*Source:* Own research.

of 0–5, where: 1 - the practice is established but has not yet been taken up, very little is happening (occurs in 1–20% of cases), 2 - the practice is only noticeable in some areas (occurs in 21–40% of cases), 3 - the practice is widely established, but not in the majority of areas (occurs in 41–60% of cases), 4 - the practice is very typical, with only some exceptions (occurs in 61–80% of cases), 5 - the practice is implemented throughout the organization, virtually without exceptions (occurs in 81–100% of cases), and 0 - the practice does not occur at all.

Based on the data contained in Table 21, it can be indicated that one-quarter of the surveyed companies have not yet implemented IoT technology. Intelligent communication and visualization systems (e.g., Andon systems) at levels 4 and 5 have been implemented by two-fifths of the surveyed companies, while one-quarter have not implemented such systems at all (level 0). Similarly, one-quarter of the surveyed steel industry companies have not yet implemented technical-business IoT solutions, and just under half have implemented these solutions at least at a moderate level (levels 3–5). RFID identification and product codes for supply chain, production, logistics, or warehousing control at levels 4 and 5 have been implemented by a total of 40.7% of the surveyed steel industry companies. The level of implementation of IoT technology enabling M2M (Machine-to-Machine) communication is slightly lower. One-third enterprises have not yet implemented this technology at all, while more than half of the organizations have implemented these solutions at least at a moderate level.

The average level of implementation of the P1 pillar in the surveyed companies reaches the level of 30–40% of activities undertaken in production processes (Table 22).

**Table 22**  Assessment of the level of P1 implementation in the steel industry enterprises – selected statistics.

| $P1$ | $\bar{x}$ | $M_e$ | $M_o$ |
|---|---|---|---|
| $q1$ | 2.47 | 3 | 0 |
| $q2$ | 2.32 | 2 | 0 |
| $q3$ | 2.56 | 3 | 0 |
| $q4$ | 2.38 | 3 | 0 |

*Source:* Own research.

The level of Big Data technology implementation in steel industry enterprises is presented in Table 23. The data indicates that these technologies are not yet widely implemented in many surveyed companies. Slightly less than one-fifth of the firms do not use technologies for tracking machine parameters and vehicle locations, one-quarter do not use software for real-time data processing and analysis, and one-third do not use management dashboards for monitoring processes and decision-making. Similarly, one-third do not use large dataset analysis to determine customer purchasing preferences and create customer profiles.

**Table 23** Assessment of the level of Big Data implementation in the steel industry enterprises (in %).

| Big Data | | Rating | | | | | |
|---|---|---|---|---|---|---|---|
| | | *0* | *1* | *2* | *3* | *4* | *5* |
| q5. | In the enterprise, machine parameters and vehicle locations are tracked to streamline processes | 17.5 | 14.6 | 8.7 | 9.7 | 23.3 | 26.2 |
| q6. | In the enterprise, software (computer system) is utilized for real-time processing and analysis of data on processes/machines/products | 25.2 | 6.8 | 12.6 | 11.7 | 24.3 | 19.4 |
| q7. | In the enterprise, managerial dashboards (providing access to current information at the managerial level) are used to processes control (monitoring) and make decisions | 31.1 | 9.7 | 13.6 | 8.7 | 18.4 | 18.4 |
| q8. | In the enterprise, data on Key Performance Indicators (KPIs) monitoring process parameters are collected on a large scale (management system) | 28.2 | 12.6 | 4.9 | 10.7 | 26.2 | 17.5 |
| q9. | In the enterprise, Big Data algorithms are employed for diagnosing production processes and machine self-diagnosis (maintenance - PM) | 33.0 | 9.7 | 9.7 | 16.5 | 17.5 | 13.6 |
| q10. | In the enterprise, large datasets are analyzed to determine customer purchasing preferences and customer profile | 33.0 | 13.6 | 6.8 | 18.4 | 17.5 | 10.7 |

*Source:* Own research.

On average, two-fifths of the surveyed steel companies report a maturity level of 4 and 5 in utilizing Big Data technologies. The most commonly implemented technologies include tracking machine parameters and vehicle locations to streamline processes, as well as monitoring KPI indicators and large-scale data collection. Slightly fewer companies use software for real-time data processing and analysis, while even fewer utilize management dashboards for monitoring processes and decision-making, Big Data algorithms for diagnosing production operations and self-diagnosing machines, and large dataset analysis to determine customer purchasing preferences and create customer profiles.

The average level of implementation of the P2 pillar in the surveyed companies reaches the level of 30–50% of activities undertaken in production procedures (Table 24)

**Table 24**   Assessment of the level of P2 implementation in the steel industry enterprises – selected statistics.

| P2 | $\bar{x}$ | $M_e$ | $M_o$ |
|---|---|---|---|
| q5 | 2.85 | 3 | 5 |
| q6 | 2.61 | 3 | 0 |
| q7 | 2.29 | 2 | 0 |
| q8 | 2.47 | 3 | 0 |
| q9 | 2.16 | 2 | 0 |
| q10 | 2.06 | 2 | 0 |

*Source:* Own research.

The data in Table 25 concerning the level of Cloud Computing technology implementation in steel industry enterprises indicates that the highest level of execution in this area pertains to cloud data storage processes. Over 50% of the surveyed firms have implemented this technology at a high level (4 and 5). However, nearly 15% of firms do not store data in the cloud at all. Two-fifths of the steel industry firms have implemented solutions providing access to various data-based services and sharing them on cloud platforms at levels 4 and 5. Nevertheless, 28.2% of the surveyed companies do not yet utilize these capabilities. Over two-fifths of the respondents use cloud data for scaling systems that support processes at a high level. Concurrently, nearly one-quarter of the firms do not use this technology at all.

The average level of implementation of the P3 pillar in the surveyed companies reaches the level of 40–80% of activities undertaken in production processes (Table 26).

**Table 25**  Assessment of the level of cloud computing implementation in the steel industry enterprises (in %).

| Cloud Computing | Rating | | | | | |
|---|---|---|---|---|---|---|
| | *0* | *1* | *2* | *3* | *4* | *5* |
| q11. In the enterprise, data is collected and stored in the cloud | 14.6 | 3.9 | 6.8 | 18.4 | 29.1 | 27.2 |
| q12. In the enterprise, access to various data-based services is provided, and these data are made available on cloud platforms | 28.2 | 3.9 | 9.7 | 15.5 | 28.2 | 14.6 |
| q13. In the enterprise, data collected and processed in the cloud are used for easy scaling of process support computer systems | 23.3 | 7.8 | 13.6 | 12.6 | 26.2 | 16.5 |

*Source:* Own research.

**Table 26**  Assessment of the level of P3 implementation in the steel industry enterprises – selected statistics.

| P3 | $\bar{x}$ | $M_e$ | $M_o$ |
|---|---|---|---|
| *q11* | 3.25 | 4 | 4 |
| *q12* | 2.55 | 3 | 0 |
| *q13* | 2.60 | 3 | 4 |

*Source:* Own research.

The level of implementation of advanced simulations in steel industry enterprises is not high, as indicated by the data presented in Table 27. The adoption of advanced simulations for production processes to improve operations varies among companies. Nearly one-third of enterprises do not use these technologies at all. However, one-fifth steel companies rate their implementation at level 3, slightly more at level 4, and one-tenth at level 5, suggesting that there is also a significant group of companies that fully or largely utilize these technologies.

Digital models (digital twins) for prototyping and process improvement are implemented at level 3 by 18.4% of the surveyed steel companies, 22.3% at level 4, and 8.7% at level 5. Simultaneously, 31.1% of companies do not use these technologies (level 0), which is an even higher percentage than in the case of production process simulations.

The average level of implementation of the P4 pillar in the surveyed companies reaches 20–40% of activities undertaken in production processes (Table 28).

**Table 27**  Assessment of the level of advanced simulation implementation in the steel industry enterprises (in %).

| *Advanced Simulation* | *Rating* | | | | | |
|---|---|---|---|---|---|---|
| | *0* | *1* | *2* | *3* | *4* | *5* |
| q14. In the enterprise, advanced computer simulations of production processes are conducted to enhance their operations | 29.1 | 8.7 | 8.7 | 19.4 | 22.3 | 11.7 |
| q15. In the enterprise, digital models (digital twin) are built using advanced simulations for prototyping and process improvement | 31.1 | 8.7 | 10.7 | 18.4 | 22.3 | 8.7 |

*Source:* Own research.

**Table 28**  Assessment of the level of P4 implementation in the steel industry enterprises – selected statistics.

| *P4* | $\bar{x}$ | $M_e$ | $M_o$ |
|---|---|---|---|
| q14 | 2.32 | 3 | 0 |
| q15 | 2.18 | 2 | 0 |

*Source:* Own research.

The level of implementation of autonomous systems in steel industry enterprises is low, as indicated by the data presented in Table 29. A total of 32.0% of organizations report no implementation of automation and robotics in production lines, while nearly the same number have applied these technologies at levels 4 and 5. Autonomous IT systems for controlling devices are not used at all by 31.1% of enterprises, and nearly one-fifth use them to a limited extent (levels 1 and 2). The implementation of these systems at levels 4 and 5 is declared by 28.2% of the surveyed enterprises.

More than 40% of the surveyed steel companies have not implemented autonomous transport scenarios. Approximately 30% use them to a limited extent. Effective use of autonomous transport systems at levels 4 and 5 is declared by 20.4% of enterprises. The use of industrial and service robots is reported by 60% of the surveyed companies. However, the level of execution of this technology varies significantly. Slightly over 17% of the surveyed companies have implemented robots at levels 1 and 2, while 30.1% rate the implementation at levels 4 or 5.

The lowest level of utilization of autonomous systems technologies is observed in the area of autonomous vehicles (AGVs) for internal transport and logistics. A total of 44.7% of enterprises rate the implementation of this technology at level 0, and only 21.4% rate it at levels 4 or 5.

The average level of implementation of the P5 pillar in the surveyed companies reaches 10–30% of activities undertaken in production processes (Table 30).

**Table 29** Assessment of the level of autonomous systems implementation in the steel industry enterprises (in %)

| Autonomous Systems | Rating | | | | | |
|---|---|---|---|---|---|---|
| | *0* | *1* | *2* | *3* | *4* | *5* |
| q16. In the enterprise, automated and robotized production lines (technological processes and/or production cells) are operational | 32.0 | 11.7 | 7.8 | 18.4 | 21.4 | 8.7 |
| q17. In the enterprise, autonomous IT-computer systems are in place, controlling their devices and responding to commands from external control systems (cloud-based applications) | 31.1 | 10.7 | 9.7 | 20.4 | 17.5 | 10.7 |
| q18. In the enterprise, transport (traffic) scenarios are developed with object and vehicle detection projects (using drones for object monitoring) | 43.7 | 12.6 | 9.7 | 13.6 | 14.6 | 5.8 |
| q19. In the enterprise, industrial and service robots are utilized in collaboration with humans (operators) | 39.8 | 8.7 | 8.7 | 12.6 | 22.3 | 7.8 |
| q20. In the enterprise, autonomous vehicles (AGVs) are used in internal transport or logistics | 44.7 | 7.8 | 8.7 | 17.5 | 16.5 | 4.9 |

*Source:* Own research.

**Table 30** Assessment of the level of P5 implementation in the steel industry enterprises – selected statistics.

| P5 | $\bar{x}$ | $M_e$ | $M_o$ |
|---|---|---|---|
| q16 | 2.11 | 2 | 0 |
| q17 | 2.15 | 2 | 0 |
| q18 | 1.60 | 1 | 0 |
| q19 | 1.92 | 2 | 0 |
| q20 | 1.68 | 1 | 0 |

*Source:* Own research.

The declared levels of implementation of universal integration technologies are summarized in Table 31. The data indicates that over one-quarter of the surveyed steel industry companies do not use computer communication and process management systems integrated with external user systems to improve communication and document flow. The remaining companies use these systems at a moderate level, with 50% at a level above 3.

**Table 31** Assessment of the level of universal integration implementation in the steel industry enterprises (in %).

| Universal Integration | Rating | | | | | |
|---|---|---|---|---|---|---|
| | 0 | 1 | 2 | 3 | 4 | 5 |
| q21. In the enterprise, computer communication and process handling systems are integrated with external user systems to streamline communication and document flow | 27.2 | 12.6 | 9.7 | 14.6 | 27.2 | 8.7 |
| q22. In the enterprise, a flexible production environment allowing real-time production corrections is operational (autonomous robots, autonomous control systems) | 24.3 | 13.6 | 10.7 | 22.3 | 19.4 | 9.7 |

*Source:* Own research.

In contrast, three-fifths of the surveyed firms utilize a flexible production environment that allows for real-time production adjustments. However, the overall level of utilization of this technology is not high. Only 30% of the surveyed companies report implementation at levels 4 and 5.

The average level of implementation of the P6 pillar in the surveyed companies reaches 30–50% of activities undertaken in production processes (Table 32).

**Table 32**  Assessment of the level of P6 implementation in the steel industry enterprises – selected statistics.

| P6 | $\bar{x}$ | $M_e$ | $M_o$ |
|---|---|---|---|
| q21 | 2.28 | 3 | 0 |
| q22 | 2.28 | 3 | 0 |

*Source:* Own research.

VR and AR technologies are not yet implemented at a high level in the steel industry. The data in Table 33 indicates that nearly half of the companies in this sector do not use VR technology for employee training or AR technology for design and service work. Approximately one-quarter of the companies have implemented VR technologies at levels 1 and 2, and one-fifth have implemented AR technologies at these levels. None of the surveyed companies have implemented these technologies at level 5.

**Table 33**  Assessment of the level of Virtual and Augmented Reality implementation in the steel industry enterprises (in %).

| *Virtual and Augmented Reality* | Rating | | | | | |
|---|---|---|---|---|---|---|
| | *0* | *1* | *2* | *3* | *4* | *5* |
| q23. In the enterprise, Virtual Reality (VR) techniques are used for employee training | 44.7 | 13.6 | 11.7 | 19.4 | 10.7 | – |
| q24. In the enterprise, Augmented Reality (AR) techniques are used for project and service work | 47.6 | 12.6 | 8.7 | 18.4 | 12.6 | – |

*Source:* Own research.

The average level of implementation of the P7 pillar in the surveyed companies reaches 10–25% of activities undertaken in production processes (Table 34).

**Table 34**  Assessment of the level of P7 implementation in the steel industry enterprises – selected statistics.

| P7 | $\bar{x}$ | $M_e$ | $M_o$ |
|---|---|---|---|
| q23 | 1.38 | 1 | 0 |
| q24 | 1.35 | 1 | 0 |

*Source:* Own research.

Based on the data presented in Table 35 concerning the level of 3D printing implementation in steel industry enterprises, it can be observed that the level of advancement in utilizing this technology varies among companies. A total of 36.9% of enterprises have not yet implemented 3D printing for designing solutions and commercial products. On the other hand, 15.5% of companies rate their implementation at level 4, and 10.7% at level 5, indicating that there is also a significant group of firms that fully or largely utilize 3D printing for product design.

The data also shows that 3D printing is used to ensure equipment continuity by printing spare parts in less than half of the enterprises. A total of 41.7% of companies do not use this technology (level 0). Implementation is rated at level 3 by 18.4% of firms, at level 4 by 12.6%, and level 5 by only 7.8%. Similarly, the level of 3D printing utilization for creating prototypes is varied. A total of 37.9% of enterprises do not use this technology (level 0), while 22.3% rate their implementation at level 4, and 10.7% at level 5, indicating that some enterprises are intensively using this technology.

**Table 35**  Assessment of the level of additive manufacturing/3D printing implementation in the steel industry enterprises (in %).

| *Additive Manufacturing/3D Printing* | *Rating* | | | | | |
|---|---|---|---|---|---|---|
| | *0* | *1* | *2* | *3* | *4* | *5* |
| q25. In the enterprise, 3D printing is used for designing solutions and products for customers (commercial product printing) | 36.9 | 10.7 | 12.6 | 13.6 | 15.5 | 10.7 |
| q26. In the enterprise, 3D printing is used to ensure equipment continuity (printing spare parts for own use) | 41.7 | 7.8 | 11.7 | 18.4 | 12.6 | 7.8 |
| q27. In the enterprise, 3D printing is used to print prototypes (models) to shorten product design time | 37.9 | 8.7 | 9.7 | 10.7 | 22.3 | 10.7 |

*Source:* Own research.

The average level of implementation of the P8 pillar in the surveyed companies reaches the level of 20–40% of activities undertaken in production processes (Table 36).

**Table 36**  Assessment of the level of P8 implementation in the steel industry enterprises – selected statistics.

| *P8* | $\bar{x}$ | $M_e$ | $M_o$ |
|---|---|---|---|
| *q25* | 1.92 | 2 | 0 |
| *q26* | 1.75 | 2 | 0 |
| *q27* | 2.03 | 2 | 0 |

*Source:* Own research.

The level of cybersecurity systems implementation in steel industry enterprises is very high compared to other Industry 4.0 technologies (Table 37). Half of the surveyed firms have implemented dedicated cybersecurity systems at levels 4 and 5, and 15% at level 3. Despite the critical importance of this technology in the digital economy, 22.3% of enterprises still do not use dedicated cybersecurity systems (level 0), and one-tenth companies use them at a minimal level (levels 1 and 2), thereby exposing themselves to the risk of cyberattacks.

The use of basic security measures, such as passwords, encryption, and codes, is significantly more common. Only 1.2% of firms do not employ these measures (level 0), indicating that almost all enterprises have implemented at least basic protections. Furthermore, 65% of companies rate their implementation of these measures at level 5, indicating a very high level of advancement and widespread use of these technologies for data protection

**Table 37**    Assessment of the level of cybersecurity implementation in the steel industry enterprises (in %).

| Cybersecurity | Rating | | | | | |
|---|---|---|---|---|---|---|
| | *0* | *1* | *2* | *3* | *4* | *5* |
| q28. In the enterprise, dedicated cybersecurity systems are employed | 22.3 | 6.8 | 3.9 | 15.5 | 25.2 | 26.2 |
| q29. In the enterprise, security measures such as passwords, encryption, and codes are implemented to enhance data security | 1.2 | 2.9 | 4.9 | 12.6 | 13.6 | 65.0 |

*Source:* Own research.

The average level of implementation of the P9 pillar in the surveyed companies reaches the level of 40–80% of activities undertaken in production processes (Table 38).

**Table 38**    Assessment of the level of P9 implementation in the steel industry enterprises – selected statistics.

| *P9* | $\bar{x}$ | $M_e$ | $M_o$ |
|---|---|---|---|
| *q28* | 2.93 | 4 | 5 |
| *q29* | 4.30 | 5 | 5 |

*Source:* Own research.

In summary, steel industry enterprises in Poland exhibit a varied level of implementation of Industry 4.0 technologies, with noticeable differences between the most advanced areas and those with lower levels of advancement (Figure 6). The most advanced areas include IoT and cloud technologies, where over 50% of the companies rate data storage in the cloud as advanced, as well as basic cybersecurity measures such as passwords, encryption, and codes, which are widely used and implemented at a high level. Technologies related to intelligent communication and visualization systems, RFID, and real-time KPI monitoring and analysis also receive relatively high implementation ratings.

Conversely, areas with lower levels of advancement include M2M communication, where the largest percentage of enterprises have not yet implemented these solutions, and advanced dedicated cybersecurity systems. 3D printing technologies, particularly for spare parts and prototypes, as well as advanced simulations and digital twins, also show a low level of implementation in a significant number of companies.

A higher level of maturity is recorded by steel companies with mixed capital compared to enterprises with Polish or foreign capital. The declared level of implementation of digital technologies is also higher in companies that have the longest market experience, have been on the market for nearly 30 years, compared to companies with a shorter presence on the market. A higher level of digital maturity is also observed in the largest enterprises (with more than 250 employees), and slightly lower in medium-sized enterprises, and the lowest in small enterprises. It is characteristic of the steel industry that regardless of the characteristics of the surveyed companies (size, capital background, length of presence on the market), the highest level of technology implementation concerns Pillar 9, which is cybersecurity. Almost all of the surveyed companies recognize the importance of this pillar and strive to implement cybersecurity technologies at the highest possible level.

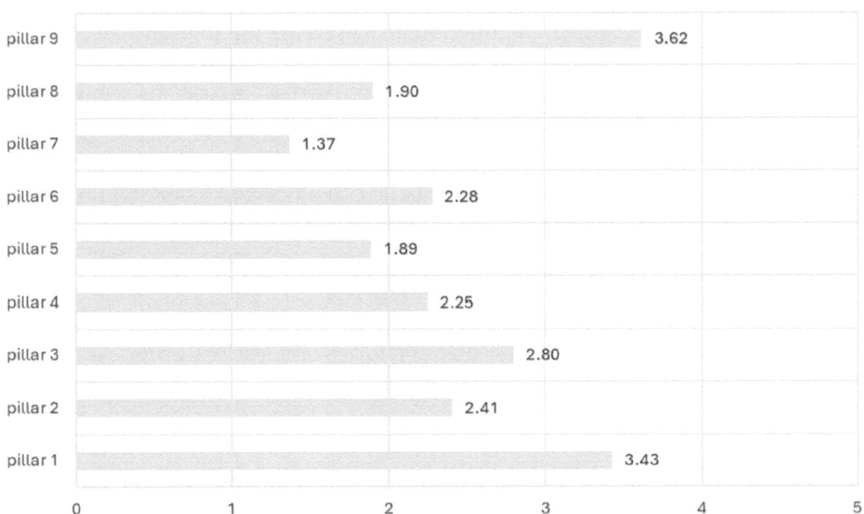


**Fig. 6** Assessment of the level of maturity of pillars of industry 4.0 in the steel industry enterprises – mean values.
*Source:* Own research.

Despite the high level of implementation in certain areas, there is considerable potential for improvement in others, indicating the need for further development, investment, education, and financial support to increase the adoption of these technologies across the steel industry. This would result in tangible benefits, such as enhanced competitiveness and innovation of enterprises in the market.

**Cluster Analysis (Steel Sector)**

The aim of the cluster analysis was to identify groups of technological factors within the nine pillars of Industry 4.0 that differ in their degree of implementation maturity in manufacturing enterprises in the steel sector. This enables a better understanding of which technologies are more or less advanced (mature) in the context of Industry 4.0 and how companies are managing their implementation. Additionally, it provides insights into how the steel industry is adapting to the challenges and opportunities arising from modern technologies.

To conduct the cluster analysis of the business maturity levels of manufacturing companies in the steel industry in the context of Industry 4.0, one of the hierarchical clustering methods—Ward's method—was employed. In this method, the distance between clusters is determined by the difference between the sums of squared deviations of individual units from the centroid of the groups to which these units belong. The distance between variables was calculated based on the squared Euclidean distance.

The cluster analysis resulted in a division into four clusters, which characterize different levels of maturity (Figure 7). Within each of these categories, the characteristics of the individual pillars of Industry 4.0 are similar in terms of the maturity assessment by respondents, whereas between the categories they differ from each other. Steel industry entrepreneurs assessing the maturity level related to, for example, cloud computing similarly rated the maturity level related to cybersecurity, while they rated the maturity level differently for, for example, the Internet of Things or advanced simulations.

Cluster 1 includes eight factors characterized by the lowest average business maturity scores. These factors are associated with advanced technologies such as VR and AR (q23, q24), selected autonomous systems (q18, q19, q20), and additive manufacturing/3D printing (q25, q26, q27). These are areas where steel industry enterprises exhibit the lowest levels of maturity.

Cluster 1 encompasses factors with the lowest average business maturity scores. The analysis indicates that manufacturing enterprises in the steel sector face the greatest challenges in implementing technologies related to VR and AR. These enterprises are not leveraging the potential of these innovations, which could be used for training and service work. The low variability of scores in this cluster suggests uniformity in the low level of technological advancement (Figure 8).

Cluster 2, encompassing four factors, presents an intermediate level of business maturity. The factors in this cluster include all technologies related to cloud computing (q11, q12, q13) and dedicated cybersecurity systems (q28). Manufacturing enterprises in the steel sector similarly rated the implementation

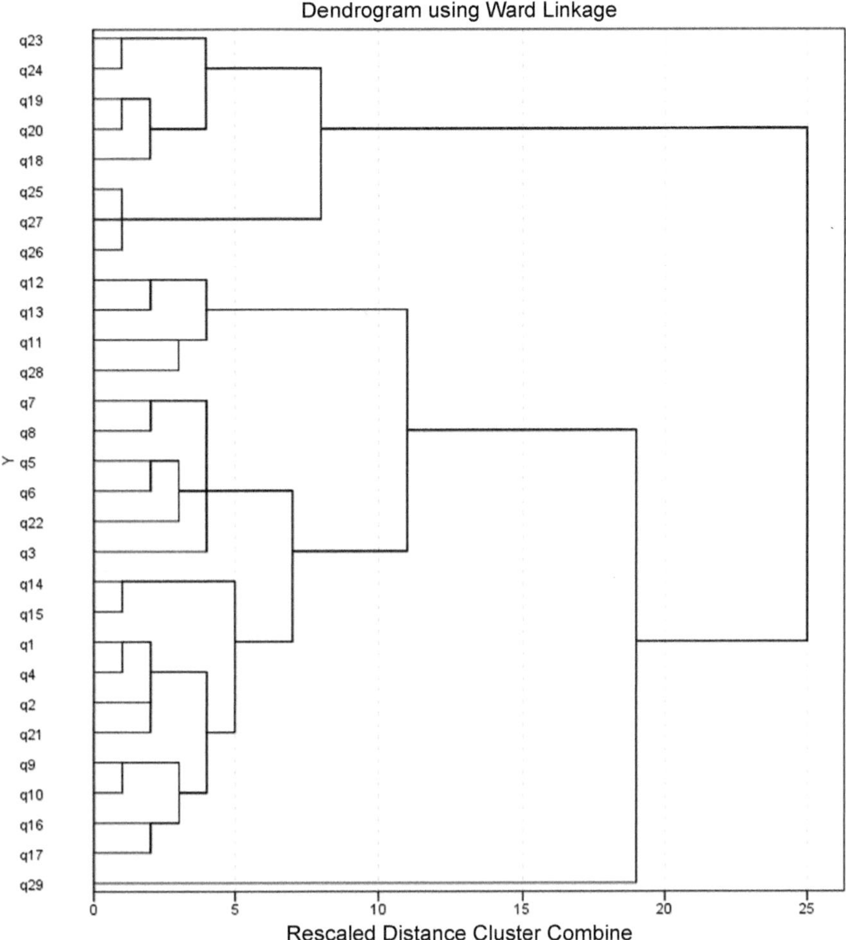



**Fig. 7** Classification of business maturity characteristics in manufacturing enterprises in the steel industry in the context of industry 4.0.
*Source:* Own research.

status of these technologies at a moderate level. The most closely rated maturity levels pertain to data-based services and the sharing of these data on various cloud platforms, as well as data in the cloud used for scaling process support IT-computer systems. The moderate variability of scores in this cluster suggests some variation in the level of maturity, but still indicates a relatively homogeneous group of variables in this area (Figure 9).

In Cluster 3, there are 16 factors characterized by high average business maturity scores in the context of Industry 4.0. The dominant factors in this cluster

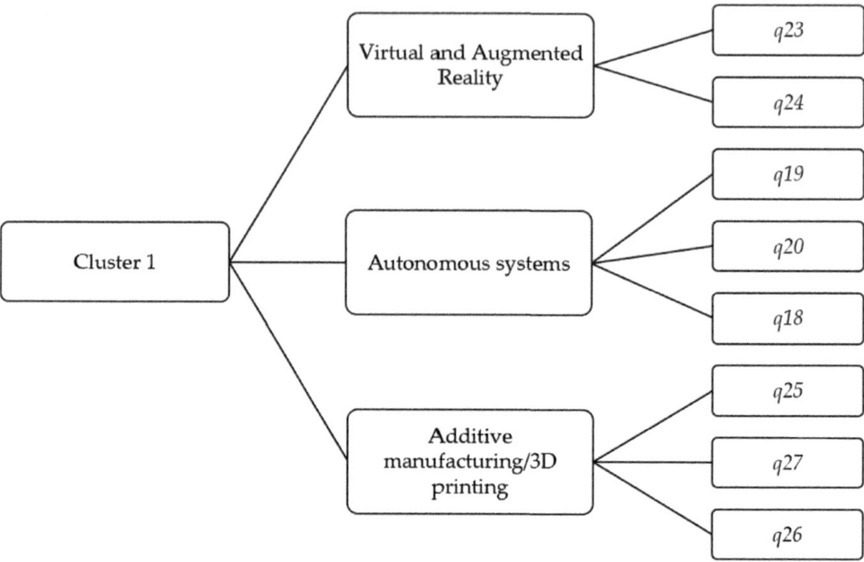

**Fig. 8** Variables identified within cluster 1 of manufacturing enterprises in the steel industry.
*Source:* Own research.

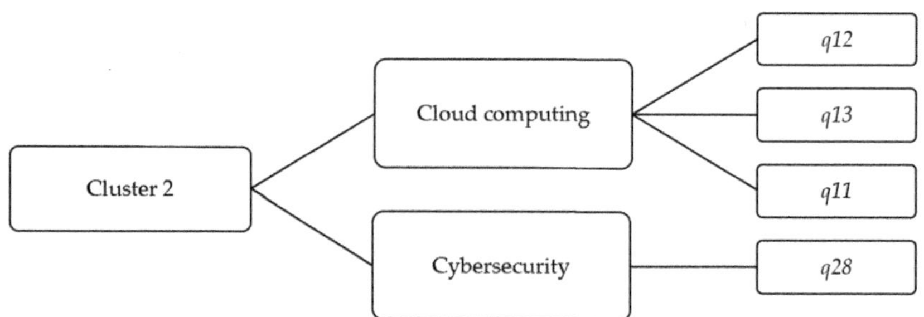

**Fig. 9** Variables identified within cluster 2 of manufacturing enterprises in the steel industry.
*Source:* Own research.

include advanced technologies such as autonomous systems (q16, q17), advanced simulation (q14, q15), universal integration (q21, q22), IoT (q1–q4), and Big Data (q5–q10).

The highest maturity within Cluster 3, which is similarly rated, is observed in technologies related to Big Data, including the tracking of machine parameters and vehicle locations to streamline processes and the use of software for real-time processing and analysis of data on processes, machines, and products. The significantly higher variability of scores in this cluster, compared to the previous clusters, suggests considerable diversity in the level of advancement of individual factors (Figure 10).

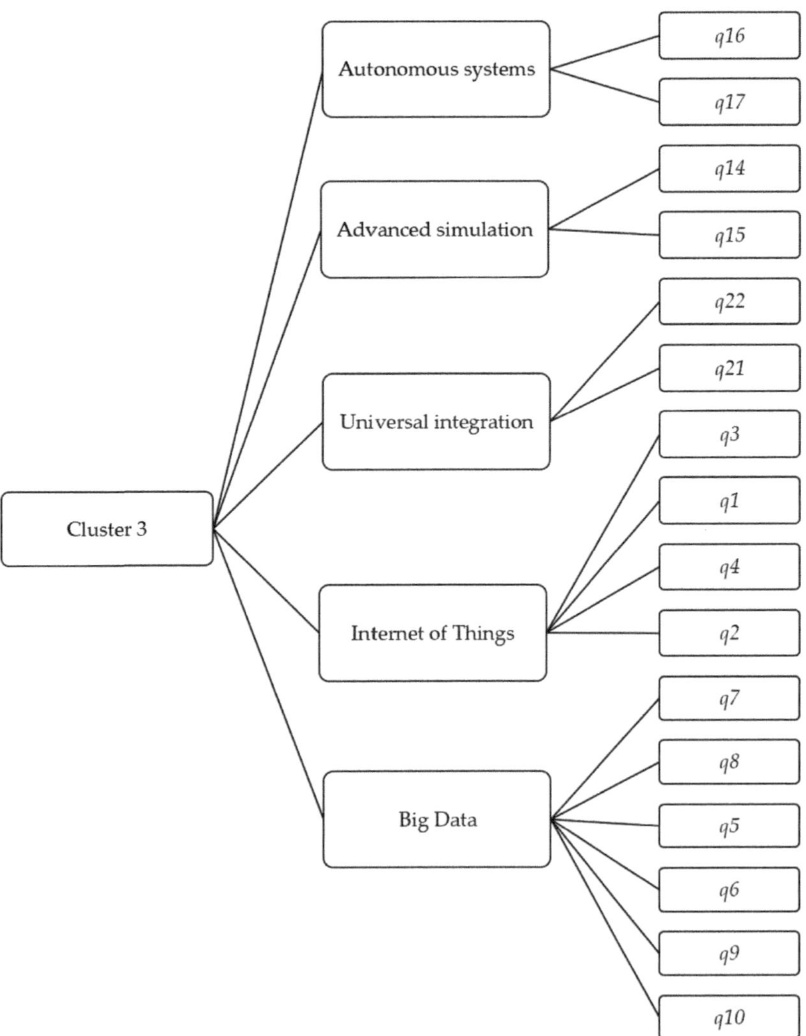


**Fig. 10** Variables identified within cluster 3 of manufacturing enterprises in the steel industry.
*Source:* Own research.

The last, fourth cluster consists of only one variable, which is the implementation of security measures such as passwords, encryption, and codes to enhance data security in enterprises (Figure 11). This indicates that this variable is sufficiently different from the others that it does not fit into any of the remaining clusters. The implementation of this technology is at the most mature level in manufacturing enterprises in the steel sector.

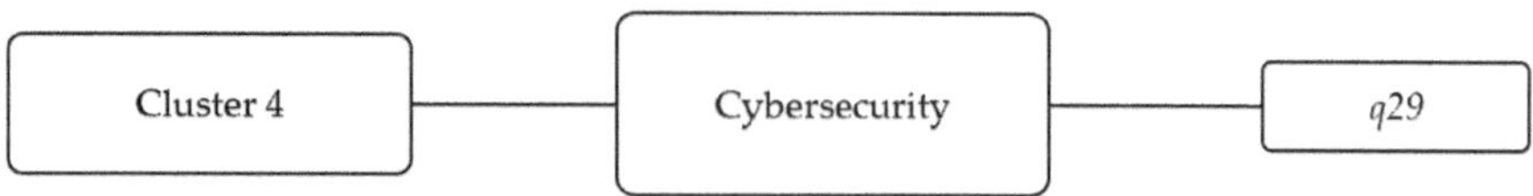


**Fig. 11**   Variable identified within cluster 4 of manufacturing enterprises in the steel industry.
*Source:* Own research.

The cluster analysis demonstrates significant differences in the assessment of the maturity of implemented technologies across the nine pillars of Industry 4.0 in manufacturing enterprises within the steel sector. The range of assessments is very wide, encompassing technologies that are rated as having low maturity in enterprises (e.g., AR) to those that are rated as highly mature (cybersecurity measures such as passwords, encryption, and codes to enhance data security).

## 3.   Maturity Models in the Food Industry

The food industry, encompassing all stages of food production, processing, distribution, and sales, is one of the most crucial sectors of the global economy. In the face of rapid technological development and increasing market demands, it is essential to understand and implement modern solutions that can significantly impact the efficiency and competitiveness of this sector. Globally, Industry 4.0 in the food sector is developing at a fast pace, with numerous examples of companies that have already implemented advanced technologies. In technologically advanced countries such as Germany, the United States, and Japan, there are many initiatives aimed at integrating modern solutions into the food value chain.

Poland, being part of the global market, also recognizes the need to adapt to these changes. Although it has strong foundations in the agri-food production sector, the implementation of Industry 4.0 technologies is still in its early stages, with numerous challenges and barriers to overcome. Understanding the level of digital maturity in the Polish food sector is crucial for the effective administration of innovative technologies. In this context, the results of conducted studies have allowed for the identification of the business maturity level of food industry enterprises in Poland.

The studies included 100 enterprises from the food sector. Among this group, 31.8% were small businesses employing 10–49 workers, 40.9% were medium-sized enterprises engaging 50–249 employees, and 27.3% were large organizations with over 250 people. Nearly one-third of these companies have been operating in the Polish market for 20 years, almost 50% have been functioning even longer, and one-quarter for over 30 years. The vast majority of the surveyed firms are enterprises with the Polish capital (Table 39).

**Table 39** Sample of the food industry enterprises (in %).

| *Items* | | % |
|---|---|---|
| Employment size | 10–49 employees | 31.8 |
| | 50–249 employees | 40.9 |
| | Above 250 employees | 27.3 |
| Year of establishment of the enterprise | Up to 1989 | 20.0 |
| | 1990–2004 | 46.4 |
| | 2005–2020 | 33.6 |
| Capital of the enterprise | Polish | 80.6 |
| | Foreign | 1.9 |
| | Mixed | 17.5 |

*Source:* Own research.

The maturity level of food industry enterprises was identified according to the conceptual and operational assumptions of the measurement model and rating scale described in Chapter 3. Table 40 presents the percentage distribution of declarations by representatives of food industry enterprises regarding the level of implementation of the Internet of Things (IoT) (pillar 1 – P1) on a scale of 0– 5, where: 1 - the practice is established but has not yet been taken up, very little is happening (occurs in 1–20% of cases), 2 - the practice is only noticeable in some areas (occurs in 21– 40% of cases), 3 - the practice is widely established, but not in the majority of areas (occurs in 41– 60% of cases), 4 - the practice is very typical, with only some exceptions (occurs in 61– 80% of cases), 5 - the practice is implemented throughout the organization, virtually without exceptions (occurs in 81– 100% of cases), and 0 - the practice does not occur at all.

The data indicates that intelligent communication, visualization, and process response systems (e.g., Andon systems) are fairly widely used, reflecting a significant degree of integration of these technologies. Nearly two-fifths of enterprises rated their implementation at level 4, indicating high advancement, while only 21.8% of enterprises have not implemented these systems at all.

The implementation of technical-business solutions based on IoT capabilities and industrial platforms indicates moderate advancement. Only one-third of the surveyed enterprises have reached level 3, while nearly one-fifth have not yet undertaken any implementations in this area.

In the context of RFID and the use of identification codes for supply chain, production, logistics, and warehousing control, there is a greater balance in the degree of technology implementation. A total of 22.7% of enterprises have reached level 4, and 20.0% have reached level 5. Therefore, these technologies are widely used and integrated into industrial operations in the surveyed food industry enterprises.

The final aspect analyzed is M2M communication using sensors and other industrial network components. In this case, 26.4% of enterprises have achieved level 4, indicating a high degree of implementation of this technology, while 22.7% of firms have not yet implemented these systems (level 0).

**Table 40** Assessment of the level of internet of things implementation in the food industry enterprises (in %).

| *Internet of Things* | *Rating* | | | | | |
|---|---|---|---|---|---|---|
| | *0* | *1* | *2* | *3* | *4* | *5* |
| q1. In the enterprise, intelligent communication, visualization, and process response systems are operational (e.g., Andon system/boards) | 21.8 | 2.7 | 18.2 | 15.5 | 39.1 | 2.7 |
| q2. In the enterprise, technical-business solutions based on the capabilities of the Internet of Things (IoT) and/or industrial platforms have been implemented | 19.1 | 5.5 | 10.0 | 31.8 | 18.2 | 15.5 |
| q3. In the enterprise, Radio Frequency Identification (RFID) and product identification codes are used to control supply chain/production/ logistics/warehouse | 18.2 | 6.4 | 12.7 | 20.0 | 22.7 | 20.0 |
| q4. In the enterprise, direct Machine-to-Machine (M2M) communication is facilitated using distributed device sensors and other components of industrial networks | 22.7 | 5.5 | 8.2 | 25.5 | 26.4 | 11.8 |

*Source:* Own research.

The average level of implementation of the P1 pillar in the surveyed companies reaches the level of 30–60% of activities undertaken in production processes (Table 41).

**Table 41** Assessment of the level of P1 implementation in the food industry enterprises – selected statistics.

| $P1$ | $\bar{x}$ | $M_e$ | $M_o$ |
|---|---|---|---|
| $q1$ | 2.55 | 3 | 4 |
| $q2$ | 2.71 | 3 | 3 |
| $q3$ | 2.83 | 3 | 4 |
| $q4$ | 2.63 | 3 | 4 |

*Source:* Own research.

The data on the implementation of Big Data technologies in food industry enterprises, presented in Table 42, indicate a varied degree of advancement in the use of these technologies across different aspects of operations. Monitoring machine parameters and vehicle locations to streamline processes is widely applied in the industry. One-third of the surveyed enterprises have implemented these technologies at level 4, and only 13.6% of businesses have not implemented these systems at all (level 0).

Real-time data processing and analysis software is also relatively frequently implemented in Polish food industry enterprises. Over 50% of the companies have implemented these solutions at levels 3 and 4. However, 20.0% of enterprises have not administered these solutions, indicating that there is still potential for further integration of these technologies.

About 60% of firms use managerial dashboards for process control and decision-making at a minimum level of 3 (including 27.3% at level 4, 25.5% at level 3, and 10% at level 5). Conversely, one in four enterprises has not implemented these tools. Large-scale KPI monitoring is employed by many organizations, with 27.3% at level 4, and 22.7% at level 3. This suggests that data collection and performance analysis are important management elements in many companies.

The implementation of Big Data algorithms for diagnosing production processes and self-diagnosis of machines is at a relatively moderate level; one-third of enterprises have implemented these algorithms at level 3, but 23.6% of firms have not used these technologies at all. The analysis of large datasets to determine customer purchasing preferences and create customer profiles is conducted to a moderate extent, with 26.4% of enterprises implementing these solutions at level 3 and 21.8% at level 4.

The average level of implementation of the P2 pillar in the surveyed companies reaches the level of 30–60% of activities undertaken in production processes (Table 43).

**Table 42** Assessment of the level of Big Data implementation in the food industry enterprises (in %).

| Big Data | Rating | | | | | |
|---|---|---|---|---|---|---|
| | *0* | *1* | *2* | *3* | *4* | *5* |
| q5. In the enterprise, machine parameters and vehicle locations are tracked to streamline processes | 13.6 | 7.3 | 8.2 | 18.2 | 34.5 | 18.2 |
| q6. In the enterprise, software (computer system) is utilized for real-time processing and analysis of data on processes/machines/products | 20.0 | 3.6 | 9.1 | 27.3 | 26.4 | 13.6 |
| q7. In the enterprise, managerial dashboards (providing access to current information at the managerial level) are used to processes control (monitoring) and make decisions | 27.3. | 4.5 | 5.5 | 25.5 | 27.3 | 10.0 |
| q8. In the enterprise, data on Key Performance Indicators (KPIs) monitoring process parameters are collected on a large scale (management system) | 15.5 | 5.5 | 11.0 | 22.7 | 27.3 | 17.3 |
| q9. In the enterprise, Big Data algorithms are employed for diagnosing production processes and machine self-diagnosis (maintenance - PM) | 23.6 | 5.5 | 10.0 | 34.5 | 17.3 | 9.1 |
| q10. In the enterprise, large datasets are analyzed to determine customer purchasing preferences and customer profile | 18.2 | 6.4 | 15.5 | 26.4 | 21.8 | 11.8 |

*Source:* Own research.

**Table 43**  Assessment of the level of P2 implementation in the food industry enterprises – selected statistics.

| $P2$ | $\bar{x}$ | $M_e$ | $M_o$ |
|---|---|---|---|
| $q5$ | 3.07 | 4 | 4 |
| $q6$ | 2.77 | 3 | 3 |
| $q7$ | 2.51 | 3 | 0 |
| $q8$ | 2.93 | 3 | 4 |
| $q9$ | 2.44 | 3 | 3 |
| $q10$ | 2.63 | 3 | 3 |

*Source:* Own research.

The analysis of data on the implementation of cloud computing technologies in food industry enterprises, presented in Table 44, indicates a varied level of advancement in the use of these technologies across different aspects. Nearly 50% of the surveyed enterprises collect and store data in the cloud at levels 4 and 5. Only 11.8% of companies have not implemented these solutions.

**Table 44**  Assessment of the level of cloud computing implementation in the food industry enterprises (in %).

| Cloud Computing | Rating | | | | | |
|---|---|---|---|---|---|---|
| | *0* | *1* | *2* | *3* | *4* | *5* |
| q11. In the enterprise, data is collected and stored in the cloud | 11.8 | 11.8 | 10.0 | 18.2 | 35.5 | 12.7 |
| q12. In the enterprise, access to various data-based services is provided, and these data are made available on cloud platforms | 19.1 | 10.0 | 10.0 | 24.5 | 27.3 | 9.1 |
| q13. In the enterprise, data collected and processed in the cloud are used for easy scaling of process support computer systems | 22.7 | 8.2 | 11.8 | 20.0 | 27.3 | 10.0 |

*Source:* Own research.

Access to various data-based services and sharing this data on cloud platforms is at a moderately high level. A total of 27.3% of enterprises have implemented these technologies at level 4, and 24.5% at level 3, indicating a significant adaptation of these solutions. However, 19.1% of enterprises do not utilize these capabilities, highlighting some gaps in the full usage of cloud computing potential.

Regarding the use of data collected and processed in the cloud for easy scaling of process-supporting systems, a similar moderate level of advancement is observed. A total of 27.3% of enterprises have implemented this technology at level 4, and 20.0% at level 3. At the same time, one-fifth enterprises have not yet implemented these technologies.

The average level of implementation of the P3 pillar in the surveyed companies reaches the level of 40–60% of activities undertaken in production processes (Table 45).

**Table 45** Assessment of the level of P3 implementation in the food industry enterprises – selected statistics.

| $P3$ | $\bar{x}$ | $M_e$ | $M_o$ |
|---|---|---|---|
| $q11$ | 2.92 | 3 | 4 |
| $q12$ | 2.58 | 3 | 4 |
| $q13$ | 2.51 | 3 | 4 |

*Source:* Own research.

The level of implementation of advanced simulations in food companies is not high. Advanced computer simulations of production processes, which are carried out in order to improve operations, have been implemented at level 4 by more than one-third of the surveyed companies from the food industry. More than one-quarter declare not to use these technologies (level 0). Digital twins using advanced simulations for prototyping and process improvement have been implemented by nearly 50% of the companies. However, the level of this administration did not exceed level 4. One-third of companies do not use such technology at all (Table 46).

**Table 46** Assessment of the level of advanced simulation implementation in the food industry enterprises (in %).

| Advanced Simulation | Rating | | | | | |
|---|---|---|---|---|---|---|
| | 0 | 1 | 2 | 3 | 4 | 5 |
| q14. In the enterprise, advanced computer simulations of production processes are conducted to enhance their operations | 27.3 | 10.0 | 9.1 | 13.6 | 35.5 | 4.5 |
| q15. In the enterprise, digital models (digital twin) are built using advanced simulations for prototyping and process improvement | 31.8 | 6.4 | 8.2 | 25.5 | 20.9 | 7.3 |

*Source:* Own research.

The average level of implementation of the P4 pillar in the surveyed companies reaches the level of 30–50% of activities undertaken in production processes (Table 47).

**Table 47** Assessment of the level of P4 implementation in the food industry enterprises – selected statistics.

| $P4$ | $\bar{x}$ | $M_e$ | $M_o$ |
|---|---|---|---|
| $q14$ | 2.34 | 3 | 4 |
| $q15$ | 2.19 | 3 | 0 |

*Source:* Own research.

The levels of execution of autonomous systems in food industry enterprises vary across the examined aspects of their operations (Table 48). For automated and robotic production lines, 33.6% of enterprises have achieved an implementation level of 4, indicating a high level of system integration. Only 9.1% of firms have not implemented these solutions (level 0).

Autonomous IT systems that control devices and respond to commands from external systems (e.g., cloud-based applications) are also significantly implemented in food industry enterprises. One-third of the enterprises have applied these systems at level 4, and 19.1% at level 3. However, 18.2% of firms have not adopted these technologies, indicating some variability in their integration.

The development of transport scenarios utilizing object and vehicle detection (e.g., drones for monitoring facilities) is less common. A total of 31.8% of enterprises have not implemented these solutions, while 21.8% of firms reported implementation at level 3, and the same proportion at level 4.

The use of industrial and service robots in collaboration with humans (operators) is relatively well-developed among the surveyed firms. One-quarter of the enterprises have implemented these technologies at level 4, and an equal proportion at level 3. However, 24.5% of firms do not use these technologies.

Autonomous Guided Vehicles (AGVs) used for internal transport and logistics are the least implemented systems among those analyzed. As many as 31.8% of enterprises do not use AGVs, and only 2.7% of firms have achieved the highest level of 5. There is significant room for development in the area of internal transport automation.

The average level of implementation of the P5 pillar in the surveyed companies reaches the level of 20–60% of activities undertaken in production processes (Table 49).

The data summarized in Table 50 indicate a moderate level of advancement in the use of universal integration technologies among food industry enterprises. Integration of communication systems and computer process management with external user systems to improve communication and document flow has been implemented at level 3 by 20% and at level 4 by 35.5% of enterprises, indicating significant adoption of these solutions. Only 16.4% of firms have not implemented these technologies.

A flexible production environment that allows for real-time production adjustments (utilizing autonomous robots and control systems) has been

**Table 48**  Assessment of the level of autonomous systems implementation in the food industry enterprises (in %).

| Autonomous Systems | Rating | | | | | |
|---|---|---|---|---|---|---|
| | 0 | 1 | 2 | 3 | 4 | 5 |
| q16. In the enterprise, automated and robotized production lines (technological processes and/or production cells) are operational | 9.1 | 6.4 | 16.4 | 21.8 | 33.6 | 12.7 |
| q17. In the enterprise, autonomous IT-computer systems are in place, controlling their devices and responding to commands from external control systems (cloud-based applications) | 18.2 | 10.0 | 10.0 | 19.1 | 33.6 | 9.1 |
| q18. In the enterprise, transport (traffic) scenarios are developed with object and vehicle detection projects (using drones for object monitoring) | 31.8 | 7.3 | 13.6 | 21.8 | 21.8 | 3.6 |
| q19. In the enterprise, industrial and service robots are utilized in collaboration with humans (operators) | 24.5 | 8.2 | 8.2 | 23.6 | 24.5 | 10.9 |
| q20. In the enterprise, autonomous vehicles (AGVs) are used in internal transport or logistics | 31.8 | 6.4 | 10.9 | 26.4 | 21.8 | 2.7 |

*Source:* Own research.

**Table 49** Assessment of the level of P5 implementation in the food industry enterprises – selected statistics.

| $P5$ | $\bar{x}$ | $M_e$ | $M_o$ |
|---|---|---|---|
| $q16$ | 3.03 | 3 | 4 |
| $q17$ | 2.67 | 3 | 4 |
| $q18$ | 2.05 | 2 | 0 |
| $q19$ | 2.48 | 3 | 0 |
| $q20$ | 2.08 | 3 | 0 |

*Source:* Own research.

implemented at level 4 by 28.2% of enterprises, and at level 3 by 24.5%. This demonstrates a substantial level of integration of these advanced technologies in the food industry.

**Table 50** Assessment of the level of universal integration implementation in the food industry enterprises (in %).

| *Universal Integration* | *Rating* | | | | | |
|---|---|---|---|---|---|---|
| | *0* | *1* | *2* | *3* | *4* | *5* |
| q21. In the enterprise, computer communication and process handling systems are integrated with external user systems to streamline communication and document flow | 16.4 | 9.1 | 9.1 | 20.9 | 35.5 | 9.1 |
| q22. In the enterprise, a flexible production environment allowing real-time production corrections is operational (autonomous robots, autonomous control systems) | 18.2 | 8.2 | 13.6 | 24.5 | 28.2 | 7.3 |

*Source:* Own research.

The average level of implementation of the P6 pillar in the surveyed companies reaches the level of 30–60% of activities undertaken in production processes (Table 51).

The analysis of data on the implementation of VR and AR technologies in food industry enterprises, presented in Table 52, indicates a relatively low level of advancement in the use of these technologies. Regarding the use of VR for employee training, 26.4% of enterprises declared a level 4, suggesting that some firms recognize the value of VR in an educational context. However, 38.2% of enterprises have not implemented these solutions at all, and no enterprises reported a level 5 implementation, highlighting that advanced use of VR in training is still rare.

**Table 51**  Assessment of the level of P6 implementation in the food industry enterprises – selected statistics.

| P6 | $\bar{x}$ | $M_e$ | $M_o$ |
|---|---|---|---|
| q21 | 2.77 | 3 | 4 |
| q22 | 2.59 | 3 | 4 |

*Source:* Own research.

A similarly low level of advancement is observed in the implementation of AR technologies for project work and services within the food industry. One-third of the businesses have reached level 3, and 14.5% have reached level 4. Moreover, nearly two-fifths of enterprises do not use AR technology at all. This indicates that there is still significant potential for the broader adoption and integration of VR and AR technologies in the food industry.

**Table 52**  Assessment of the level of Virtual and Augmented Reality implementation in the food industry enterprises (in %).

| *Virtual and Augmented Reality* | Rating | | | | | |
|---|---|---|---|---|---|---|
| | *0* | *1* | *2* | *3* | *4* | *5* |
| q23. In the enterprise, Virtual Reality (VR) techniques are used for employee training | 38.2 | 10.9 | 10.0 | 14.5 | 26.4 | – |
| q24. In the enterprise, Augmented Reality (AR) techniques are used for project and service work | 38.2 | 7.3 | 13.6 | 23.6 | 14.5 | 2.7 |

*Source:* Own research.

The average level of implementation of the P7 pillar in the surveyed companies reaches the level of 1–40% of activities undertaken in production processes (Table 53).

**Table 53**  Assessment of the level of P7 virtual/augmented reality implementation in the food industry enterprises – selected statistics.

| P7 | $\bar{x}$ | $M_e$ | $M_o$ |
|---|---|---|---|
| q23 | 1.80 | 2 | 0 |
| q24 | 1.77 | 2 | 0 |

*Source:* Own research.

Data on the implementation of 3D printing technology in food enterprises, presented in Table 54, indicate a relatively low level of advancement in the use of these technologies in the activities of this industry. The use of 3D printing to design

**Table 54** Assessment of the level of additive manufacturing/3D printing implementation in the food industry enterprises (in %).

| Additive Manufacturing/3D Printing | Rating | | | | | |
|---|---|---|---|---|---|---|
| | 0 | 1 | 2 | 3 | 4 | 5 |
| q25. In the enterprise, 3D printing is used for designing solutions and products for customers (commercial product printing) | 47.3 | 8.2 | 7.3 | 13.6 | 20.0 | 3.6 |
| q26. In the enterprise, 3D printing is used to ensure equipment continuity (printing spare parts for own use) | 47.3 | 8.2 | 7.3 | 18.2 | 16.4 | 2.7 |
| q27. In the enterprise, 3D printing is used to print prototypes (models) to shorten product design time | 49.1 | 7.3 | 8.2 | 17.3 | 13.6 | 4.5 |

*Source:* Own research.

solutions and products for customers (printing of commercial products) at level 4 is declared by every fifth company. At the same time, as many as 47.3% of companies have not yet implemented these solutions. Only 3.6% of companies have reached implementation level 5, highlighting that advanced use of 3D printing is still rare. When using 3D printing to ensure the continuity of equipment (printing spare parts for your own use), the situation is similar. 18.2% of companies implemented the technology at Level 3 and 16.4% at Level 4, suggesting that some companies are appreciating the benefits of 3D printing in the context of maintenance. However, 47.3% of companies do not use these technologies. The use of 3D printing to print prototypes to reduce product design time also shows a low level of sophistication. 17.3% of enterprises achieved implementation level 3 and 13.6% achieved level 4, suggesting moderate interest in this technology. Almost 50% of companies do not use 3D printing in this context.

The average level of implementation of the P8 pillar in the surveyed companies reaches the level of 1–30% of activities undertaken in production processes (Table 55).

**Table 55**    Assessment of the level of P8 implementation in the food industry enterprises – selected statistics.

| P8 | $\bar{x}$ | $M_e$ | $M_o$ |
|---|---|---|---|
| q25 | 1.62 | 1 | 0 |
| q26 | 1.56 | 1 | 0 |
| q27 | 1.52 | 1 | 0 |

*Source:* Own research.

The analysis of data on the level of cybersecurity systems implementation in food industry enterprises, as presented in Table 56, indicates a relatively high level of advancement in this area. Regarding the use of dedicated cybersecurity systems, 31.8% of enterprises are at level 4 of digital maturity, and 19.1% are at level 5. This indicates that a significant number of firms prioritize cybersecurity. Only 8.2% of enterprises have not implemented any cybersecurity systems.

In terms of implementing security measures such as passwords, encryption, and codes to enhance data security, 33.6% of enterprises have implemented these technologies at level 5, and 23.6% at level 4. This further emphasizes that food industry enterprises are actively adopting advanced security measures to protect their data. The absence of enterprises reporting level 0 suggests that basic security measures are at least minimally applied in all surveyed firms.

**Table 56** Assessment of the level of cybersecurity implementation in the food industry enterprises (in %).

| Cybersecurity | Rating | | | | | |
|---|---|---|---|---|---|---|
| | *0* | *1* | *2* | *3* | *4* | *5* |
| q28. In the enterprise, dedicated cybersecurity systems are employed | 8.2 | 10.9 | 11.8 | 18.2 | 31.8 | 19.1 |
| q29. In the enterprise, security measures such as passwords, encryption, and codes are implemented to enhance data security | – | 9.1 | 11.8 | 21.8 | 23.6 | 33.6 |

*Source:* Own research.

The average level of implementation of the P9 pillar in the surveyed companies reaches the level of 50–70% of activities undertaken in production processes (Table 57).

**Table 57** Assessment of the level of P9 (cybersecurity) implementation in the food industry enterprises – selected statistics.

| *P9* | $\bar{x}$ | $M_e$ | $M_o$ |
|---|---|---|---|
| *q28* | 3.12 | 4 | 4 |
| *q29* | 3.61 | 4 | 5 |

*Source:* Own research.

In summary, food industry enterprises exhibit a high level of digital maturity in the area of Industry 4.0 with regard to the implementation of cybersecurity systems, intelligent communication systems, and RFID, reflecting their awareness of the importance of data protection and operational efficiency. Notable successes have also been observed in monitoring machine parameters and vehicle locations through Big Data technology, as well as in data storage and management using Cloud Computing.

However, there remains significant potential for further development in areas such as advanced simulations, autonomous systems, flexible production systems, VR and AR, and 3D printing, which show relatively low levels of implementation in the surveyed enterprises. Continued investment and education are necessary to fully leverage the opportunities offered by these technologies, which could enhance the efficiency, innovation, and competitiveness of the food sector.

When comparing the average levels of implementation of various Industry 4.0 pillars, the greatest advancement is observed in pillar 9 (cybersecurity), while the least advancement is seen in pillar 8 (additive manufacturing) (Figure 12).

A higher level of maturity is recorded by enterprises in the food industry with mixed capital compared to enterprises with Polish or foreign capital. The declared level of digital adoption is also higher in companies that have had market experience since at least 2005, compared to companies with a longer market presence. A higher

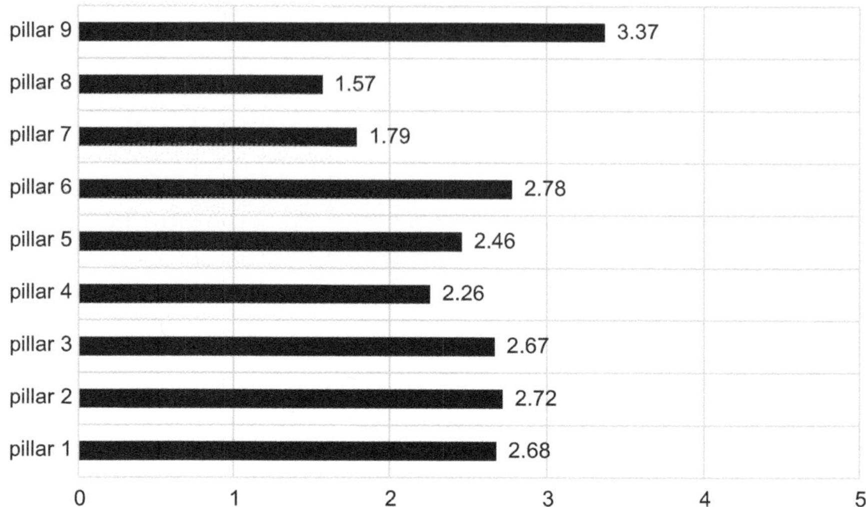


**Fig. 12**    Assessment of the level of maturity of pillars 1–9 of industry 4.0 in the food industry enterprises – mean values.
*Source:* Own research.

level of digital maturity is also observed in the largest enterprises (with more than 250 employees), and slightly lower in medium-sized enterprises, and the lowest in small enterprises. It is characteristic of the food industry that regardless of the features of the surveyed companies (size, capital origin, length of presence on the market), the highest level of technology implementation is seen in pillar 9, which is cybersecurity. Almost all of the surveyed companies recognize the importance of this pillar and strive to implement cybersecurity technologies at the highest possible level.

**Cluster Analysis (Food Industry)**

The cluster analysis aimed to identify groups of technological factors within the nine pillars of Industry 4.0 that differ in their degree of implementation maturity in manufacturing enterprises within the food sector. The results of the analysis provide a better understanding of which technologies are more or less advanced (mature) in the context of Industry 4.0 and how manufacturing enterprises are managing their implementation.

To conduct the cluster analysis of the business maturity levels of manufacturing enterprises in the food industry in the context of Industry 4.0, hierarchical Ward's clustering method was employed. In this method, the distance between clusters is defined as the difference between the sums of squared deviations of individual units from the centroid of the groups to which these points belong. The distance between variables was determined based on the squared Euclidean distance.

Based on the analysis conducted for the assessments of the various pillars of Industry 4.0, three clusters were identified (Figure 13). Within each of these three categories, the characteristics of the individual pillars of Industry 4.0 are similar in

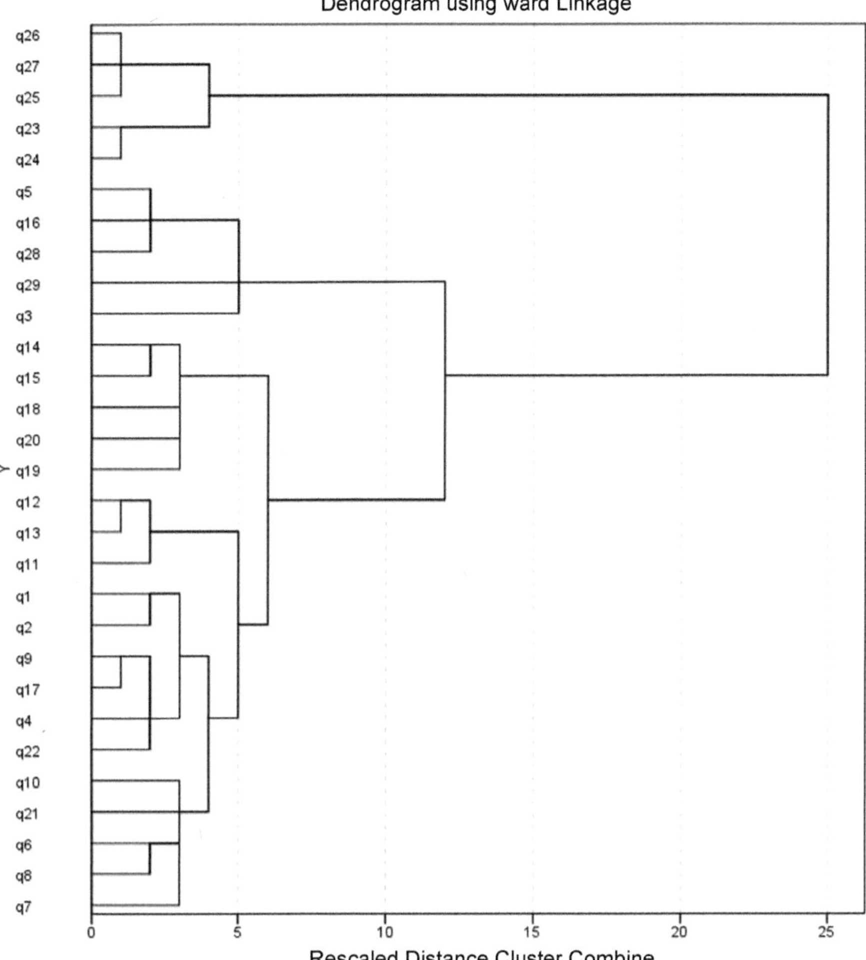


**Fig. 13** Classification of business maturity characteristics in manufacturing enterprises in the food industry in the context of industry 4.0.
*Source:* Own research.

terms of the maturity assessment by respondents, whereas between the categories they differ from each other. Entrepreneurs in the food industry evaluating the maturity level related to, for example, cloud computing or advanced simulation similarly rated the maturity level related to, for example, universal integration, whereas they rated the maturity level differently for, for example, VR/AR or additive manufacturing/3D printing.

Cluster 1 encompasses five factors that are characterized by decidedly the lowest average business maturity scores. These factors are associated with advanced technologies such as Virtual and Augmented Reality (q23, q24) and additive manufacturing/3D printing (q25, q26, q27). These are areas, particularly additive manufacturing/3D printing, where manufacturing enterprises in the food industry exhibit the lowest levels of maturity and even face difficulties in implementing these technologies. Cluster 1 is characterized by low standard deviation and low interquartile range, indicating uniformity in the low level of business maturity (Figure 14).

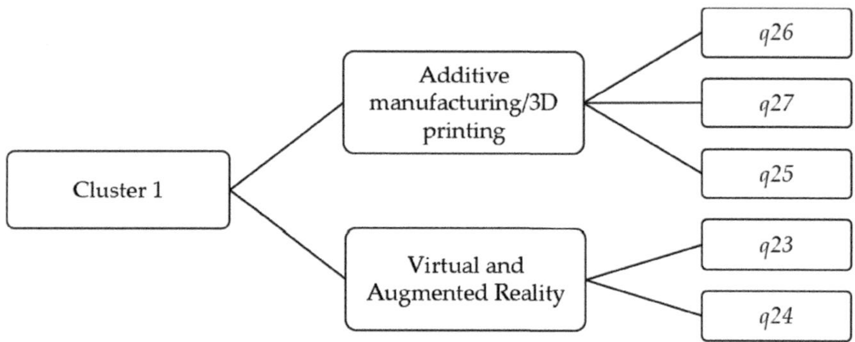


**Figure 14.** Variables identified within cluster 1 of manufacturing enterprises in the food industry. *Source:* Own research.

Cluster 2, which encompasses as many as 19 factors, is characterized by average business maturity scores in the context of Industry 4.0 for the food industry. The dominant factors in this cluster include technologies such as advanced simulation (q14, q15), autonomous systems (q17– q20), cloud computing (q11– q13), universal integration (q21, q22), Big Data (q6– q10), and IoT (q1, q2, q4). In these areas, food industry enterprises achieve a moderate level of maturity. The high variability in the scores of the factors within this cluster suggests significant diversity in the level of advancement of the individual factors (Figure 15).

Cluster 3, encompassing five factors, presents the highest level of business maturity. The factors in this cluster include all technologies related to cybersecurity (q28, q29), as well as one technology each from Big Data (q5), the IoT (q3), and autonomous systems (q16). Food industry manufacturing enterprises similarly rated the implementation status of these technologies at a relatively high level. The most closely aligned maturity ratings pertain to the tracking of machine parameters and vehicle locations to streamline processes, as well as the operation of automated and robotized production lines. The moderate variability of scores in this cluster suggests slight variability in the level of maturity, but still indicates a relatively homogeneous group of variables in this area (Figure 16).

The results of the analysis indicate significant differences in the business maturity ratings between the clusters. The range of ratings is very broad, encompassing technologies that are rated as having low maturity in enterprises (e.g., using 3D printing to print prototypes to shorten product design time) and those that are rated as highly mature (e.g., cybersecurity measures such as passwords, encryption, and codes to enhance data security).

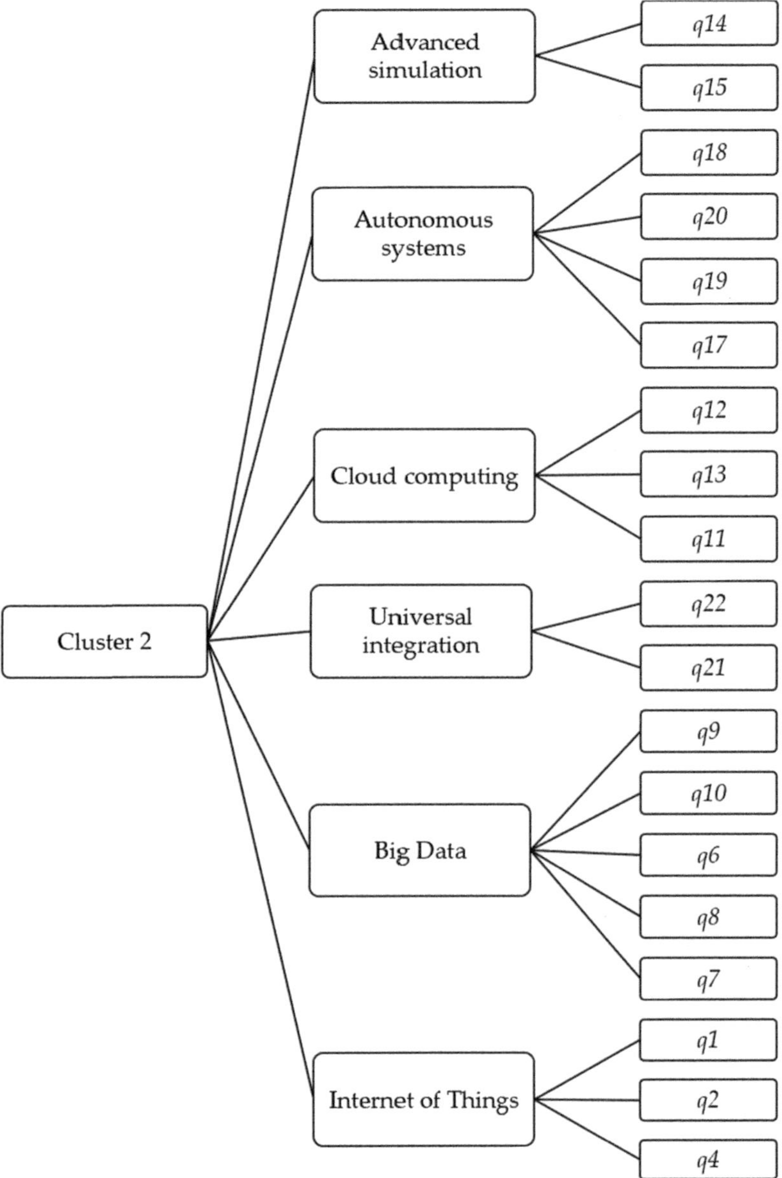


**Fig. 15** Variables identified within cluster 2 of manufacturing enterprises in the food industry.

*Source:* Own research.

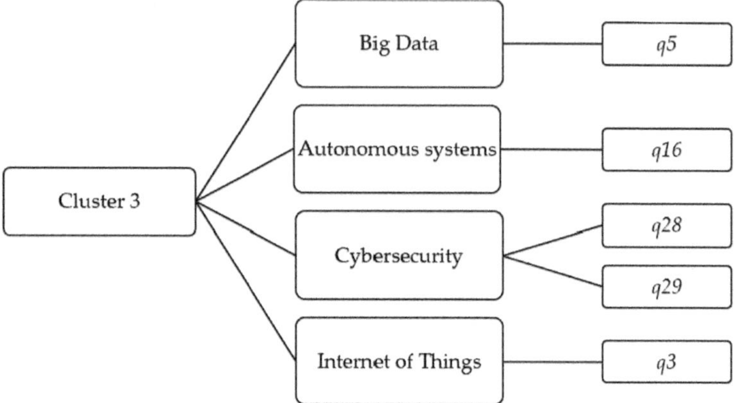


**Fig. 16** Variables identified within cluster 3 of manufacturing enterprises in the food industry. *Source:* Own research.

The research conducted on Polish enterprises from the automotive, steel, and food industries revealed significant insights into the implementation levels of Industry 4.0 technologies (level of business maturity models). The findings indicate varied levels of technological maturity across different sectors and pillars of Industry 4.0.

In the automotive industry, high levels of implementation were noted in IoT, Big Data, cloud technologies, advanced simulations, and cybersecurity systems. Over 82.7% of companies rated intelligent communication systems at level 3 or higher, demonstrating widespread use. The integration of IoT and RFID technologies is well established, with significant portions of enterprises utilizing these technologies to enhance operational efficiency. However, there is room for improvement in areas such as autonomous guided vehicles (AGVs) and VR/AR technologies.

The steel industry showed substantial implementation of IoT and Cloud Computing technologies, particularly in data storage and security measures. 40.7% of companies implemented RFID technologies at high levels for supply chain and logistics management. Advanced simulations and digital twins were less commonly used, indicating potential areas for growth. The use of autonomous systems and 3D printing technologies was limited, highlighting significant opportunities for further development.

The food sector exhibited moderate advancement in IoT and Big Data technologies. Cloud Computing was widely adopted, with nearly half of the companies using these solutions at advanced levels. However, advanced simulations and digital twins were less frequently implemented, with many companies yet to adopt these technologies. The implementation of VR and AR remained low, suggesting a need for increased focus on these areas.

# 5 | Conclusion

The research presented in the book is a comprehensive analysis of the business maturity of Polish manufacturing companies within the context of Industry 4.0. The findings of this study highlight the critical drivers of technological advancements and offer insights into different levels of business maturity across sectors.

The study conceptualizes and operationalizes a research model aimed at assessing the business maturity of industrial enterprises. This model is based on the principles of Industry 4.0, covering nine key pillars: Internet of Things (IoT), Big Data and Analytics, Cloud Computing, Advanced Simulation, Autonomous Systems, Universal Integration, Virtual and Augmented Reality, Additive Manufacturing, and Cybersecurity.

The model defines business maturity as an organization's ability to adapt, change, and grow in a dynamic market environment. This includes improving processes, structures, and resources to achieve strategic and operational goals. The study shows that achieving high business maturity requires a holistic approach that integrates technological, process, and organizational dimensions.

The methodological approach adopted in this study includes the construction of a tool for measuring business maturity and verifying its reliability. The study used a detailed survey covering the nine pillars of Industry 4.0, which was then used to assess the business maturity of Polish manufacturing companies. The results indicate the high reliability of the measurement scale, confirming its effectiveness in assessing the technological readiness and business maturity of enterprises.

The research presented in this book assesses the maturity levels of the nine pillars of Industry 4.0 in various sectors of the Polish manufacturing industry. The study used average values to quantify maturity levels and highlighted significant differences between the automotive, steel, and food industries. See Table 1 for a summary of this information.

The assessment reveals significant differences in the maturity levels of Industry 4.0 pillars in the automotive, steel, and food industries. The automotive industry has consistently demonstrated a higher level of maturity, reflecting the advanced integration and use of Industry 4.0 technologies. The sector's focus on

**Table 1** Comparison of the level of maturity of Industry 4.0 pillars in the analyzed industries.

| Pillar | Automotive | Steel Industry | Food Industry |
|---|---|---|---|
| Internet of Things (IoT) | High maturity with an average value of 4.2. The automotive sector has widely adopted IoT for real-time monitoring and predictive maintenance, leading to the optimization of manufacturing processes. | Moderate incidence with an average value of 3.5. IoT is used to monitor and control production processes, but integration is not as common as in the automotive sector. | Lower maturity with an average value of 2.8. The food industry is lagging behind in terms of IoT implementation, mainly using it for basic tracking and quality control. |
| Big Data | High maturity with an average value of 4.0. Automotive companies use Big Data analytics to improve product quality and customer satisfaction through detailed data analysis. | Moderate incidence with an average value of 3.2. Big Data is used to optimize processes and manage energy, but not to the same extent as in the automotive sector. | Lower maturity with an average value of 2.5. The food industry uses Big Data primarily for supply chain management and basic market analysis. |
| Cloud Computing | High maturity with an average value of 4.1. Cloud Computing is used to store, process, and collaborate data across global manufacturing networks. | Moderate incidence with an average value of 3.4. Cloud solutions are used to increase operational efficiency, but they face resistance due to security concerns. | Lower maturity with an average value of 2.7. The use of Cloud Computing is limited, mainly to basic data storage and sharing. |
| Advanced Simulation | High incidence with an average value of 4.3. Advanced simulation technologies are widely used for virtual prototyping and testing, reducing time-to-market. | Moderate incidence with an average value of 3.6. Simulation tools are used to optimize processes and analyze failures. | Lower maturity with an average value of 2.9. Simulation is used to a lesser extent, mainly for process improvement and new product development. |
| Autonomous Systems | High maturity with an average value of 4.4. The use of autonomous systems, including robotics and automated guided vehicles (AGVs), is common in manufacturing and logistics. | Moderate incidence with an average value of 3.7. Autonomous systems are used for specific tasks, such as material handling and quality control. | Lower maturity with an average value of 3.0. The use of autonomous systems is limited to automated packaging and sorting processes. |

*Contd.*

**Table 1** *Contd.*

| Pillar | Automotive | Steel Industry | Food Industry |
|---|---|---|---|
| Universal Integration | High maturity with an average value of 4.2. The automotive sector is at the forefront of integrating various technologies and digital systems throughout the supply chain. | Moderate incidence with an average value of 3.3. Integration work is underway, but legacy systems pose a challenge. | Lower maturity with an average value of 2.8. The integration is mainly focused on traceability and supply chain management. |
| Virtual and Augmented Reality (VR/AR) | High incidence with an average value of 4.1. VR and AR are widely used for design, training, and maintenance purposes. | Moderate incidence with an average value of 3.4. VR and AR are used for training and remote assistance. | Lower incidence with an average value of 2.6. VR and AR are at an early stage of implementation, mainly in training and marketing. |
| Additive Manufacturing | High maturity with an average value of 4.0. Additive manufacturing is used for prototyping and small-scale production of complex parts. | Moderate incidence with an average value of 3.2. Additive manufacturing is primarily used for the production of tools and dies. | Lower maturity with an average value of 2.7. Adoption is limited to niche applications such as custom food products and packaging. |
| Cybersecurity | High incidence with an average value of 4.3. Robust cybersecurity measures are implemented to protect critical data and systems. | Moderate incidence with an average value of 3.5. Cybersecurity is a growing concern, and efforts are focused on protecting industrial control systems. | Moderate incidence with an average value of 3.5. Cybersecurity is a growing concern, and efforts are focused on protecting industrial control systems. |

*Source:* Own preparation.

innovation, efficiency, and competitiveness makes it a leader in the adoption of new technologies.

The automotive industry is inherently technology-driven, with a constant need for innovation to remain competitive. Vehicle manufacturing involves complex processes that benefit greatly from advances in automation, IoT, and data analytics. Automotive companies have historically invested in advanced manufacturing techniques such as robotics and Computer-Aided Design (CAD), which naturally incorporate newer Industry 4.0 technologies such as advanced simulations and autonomous systems.

The automotive industry usually has more financial resources to invest in new technologies. Large car manufacturers can allocate significant budgets to R&D (research and development) and technological upgrades. The scale of production in the automotive industry allows for better amortization of technological investments in relation to the larger number of units produced, making advanced technologies more economically viable.

The automotive industry has a strong culture of innovation, driven by intense competition and consumer demand for advanced features. This culture supports the rapid adoption of new technologies. The industry attracts a highly skilled workforce, including engineers and technicians, who are able to implement and manage advanced technologies.

Consumers in the automotive market have high expectations for innovation, safety, and performance. This prompts manufacturers to continuously implement cutting-edge technologies to meet market demands. The global nature of the automotive market means that companies have to compete internationally, forcing them to adopt Industry 4.0 technologies in order to remain competitive.

The steel industry shows a moderate level of maturity, with a focus on process optimization and energy management. While the industry is making strides in digital transformation, it faces challenges such as legacy systems and security concerns. The steel industry focuses on optimizing manufacturing processes. While it uses technologies like IoT and Big Data to monitor and control processes, the pace of adoption is slower due to the industry's reliance on heavy, capital-intensive equipment. The presence of legacy systems and equipment in steel plants makes the integration of new technologies more difficult and costly.

Steel production requires significant capital investment in manufacturing technologies and infrastructure. The high cost of retrofitting these facilities with new technologies can be a deterrent. The steel industry often faces market volatility in terms of raw material prices and demand, which can affect the availability of funds for technological progresses.

The steel industry often has well-established operating procedures and practices. This can cause resistance to change and slow down the adoption of new technologies. There may be a skills' gap among the workforce, with the need for training programs to upskill employees in digital technologies and data analytics.

The steel industry primarily serves industrial customers who require reliability and efficiency. While there is a demand for innovation, the pressure to adopt new technologies is less direct compared to consumer-facing industries. The cyclical nature of the demand for steel products influences investment decisions. During a downturn, companies may postpone technology investments in order to conserve resources.

The food industry is lagging behind in most pillars, indicating a slower adoption of Industry 4.0 technologies. The sector's lower maturity level is attributed to limited resources, greater cost sensitivity, and a slower pace of technological change. The food industry generally involves less high-tech processes compared to the automotive and steel industries. This results in a slower adoption of advanced digital technologies. High product variability and stringent food safety regulations require specific technological solutions that may not always be compatible with general Industry 4.0 applications, making the implementation process more complex.

The food industry operates on lower profit margins compared to the automotive and steel industries. This limits the financial resources available for investments in new technologies. Due to the high price sensitivity among consumers, food companies may prioritize cost-cutting and efficiency improvements over large-scale technology investments.

Many companies in the food industry rely on traditional production practices. This conservative approach can make it difficult to implement new, unfamiliar technologies. The need to comply with stringent food safety regulations can also detract from technological innovation.

The food industry needs to adapt to rapidly changing consumer preferences and trends. While this drives some innovation, the focus is often on product development rather than optimizing the manufacturing process.

Differences in maturity levels highlight the need to develop tailored strategies to support the digital transformation of each sector. Automotive companies should continue to innovate and integrate new technologies, steel companies should focus on overcoming the challenges of legacy systems, and food companies must prioritize investments in key technologies to catch up. Policymakers and industry associations can play a key role in providing the necessary support and incentives to facilitate this transition.

A comparison of the three analyzed industries is shown in Figure 1. This figure presents an assessment of maturity levels across the nine key pillars of Industry 4.0 in three industry sectors: automotive, steel, and food. The data are represented by averages, illustrating the average maturity level for each pillar in each industry sector.

**Automotive industry:**   This sector shows the highest level of maturity in most pillars compared to other sectors. Maturity levels consistently exceed 3.0 for pillars such as cloud computing, advanced simulations, autonomous systems, and cybersecurity, with the highest level in cybersecurity at 3.78.

**Steel industry:** This sector shows diverse levels of maturity, with the highest in cyber solutions and security at 3.62 and the lowest in universal integration at

1.37. Maturity levels for most pillars are around 2.0, indicating a moderate level of sophistication.

**Food industry:** This sector shows a lower level of maturity overall, with peaks in cybersecurity at 3.37 and lows in universal integration at 1.57. The maturity of the other pillars remains at around 2.0, suggesting an early stage of Industry 4.0 adoption.

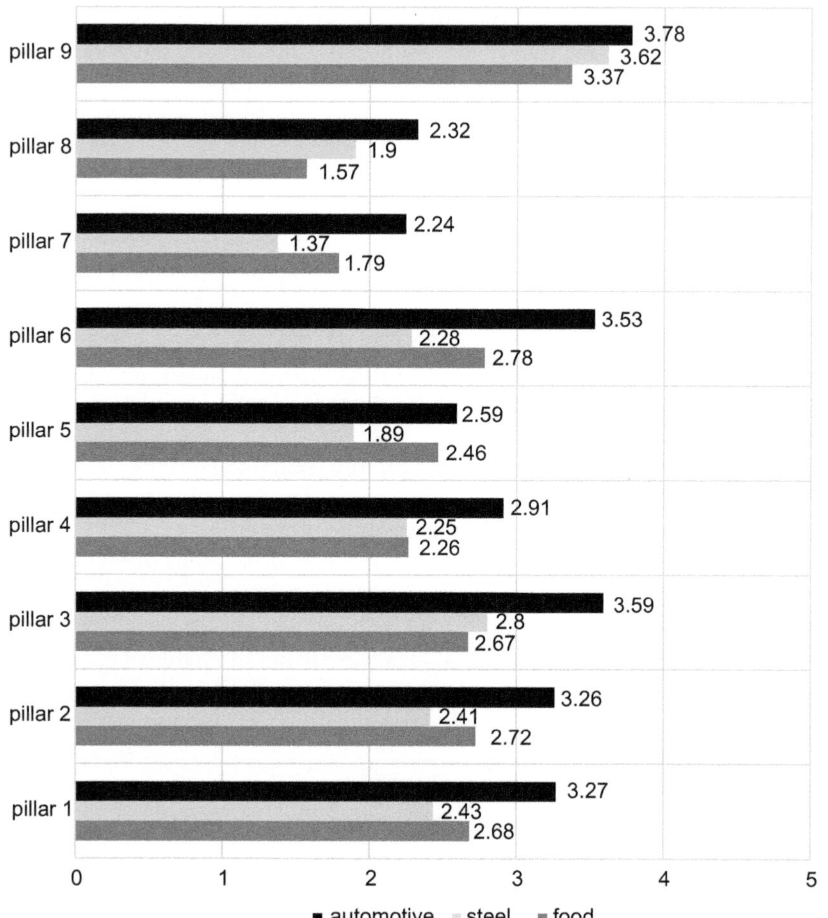


**Fig. 1** Assessment of the maturity level of the 9 pillars of Industry 4.0 in industries: Average values.
*Source:* Own research.

The results indicate (Figure 1) that the automotive industry is leading the way in Industry 4.0 maturity, followed by the steel and food industries.

In the automotive industry, maturity levels in all pillars are consistently higher compared to the other two sectors. This indicates a significant advantage in the implementation of Industry 4.0 technologies in the automotive sector. The pillars of cloud computing, advanced simulation, autonomous systems, and cybersecurity stand out with average maturity levels above 3.0, reflecting a high degree of technological integration. Cybersecurity, in particular, peaks at 3.78, demonstrating the industry's strong focus on securing technological advancements and data.

The steel industry has a more diverse maturity landscape. Although it does not reach as high a level as the automotive industry, it shows significant progress in cybersecurity, with an average maturity level of 3.62. This suggests that even if other pillars such as universal integration (1.37 average) and big data analytics may be lagging behind, there is a strong focus on securing technology deployments. Most of the other pillars hover around the 2.0 level, indicating moderate adoption and technological integration. This pattern reveals the steel industry's selective focus on some key areas, while other aspects of Industry 4.0 remain less developed.

The food industry, on the other hand, shows an overall lower level of maturity in all pillars of Industry 4.0. Cybersecurity is again emerging as the most advanced pillar with an average level of 3.37, similar to the trends seen in the automotive and steel industries. However, other pillars, such as universal integration, which averages 1.57, and big data analytics, remain underdeveloped, with a maturity level of around 2.0. This suggests that the food and beverage industry is in the early stages of Industry 4.0 adoption, focusing primarily on securing its technology implementations but facing challenges related to wider integration and adoption of advanced technologies.

Figure 2 illustrates the uneven implementation of Industry 4.0 technologies in the analyzed sectors. The automotive industry is at the forefront of the end-to-end integration of these technologies, particularly excelling in cybersecurity and other advanced pillars. The steel industry exhibits a more selective deployment pattern, with strong safety measures but slower integration in other areas. The food industry, while prioritizing cybersecurity, is lagging behind in terms of overall technology adoption, highlighting significant room for growth and development.

This comparative analysis highlights the varying pace and focus of Industry 4.0 adoption across sectors. It emphasizes the automotive industry's leadership role, setting the standard for technological integration. Meanwhile, the steel and food industries show potential for targeted improvements, especially in achieving a more sustainable and comprehensive adoption of Industry 4.0 pillars. Such insights are critical to future investment strategies and efforts to strengthen the technological capabilities of these industries, ensuring more even progress in the digital transformation era.

Table 2 provides a detailed breakdown of the maturity levels of the Industry 4.0 pillars, categorized according to different characteristics of industrial enterprises, such as employment size, year of establishment, and origin of capital. Maturity levels are expressed in average values.

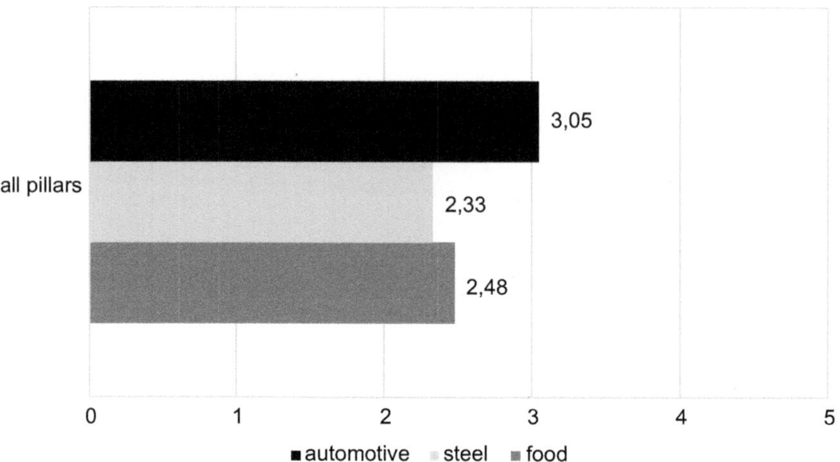


**Fig. 2**  Assessment of the maturity level of all pillars of Industry 4.0 in industries: Average values.
*Source:* Own research.

Analyzing the data by employment size, the largest enterprises (with more than 250 employees) show the highest level of maturity, with averages ranging from 3.24 to 3.34. This suggests that larger companies have more resources and opportunities to adopt advanced Industry 4.0 technologies. Medium-sized enterprises (50–249 employees) show a moderate level of maturity, averaging between 2.72 to 3.12, reflecting a reasonable but less comprehensive adoption of these technologies. Small enterprises (10–49 employees) show the least maturity, with average values around 1.42 to 2.74, indicating significant challenges related to the implementation and integration of Industry 4.0 technologies.

The year of establishment also plays an important role in the maturity levels of the Industry 4.0 pillars. Companies founded before 1989 generally show a higher level of maturity, with average values ranging from 2.02 to 3.14, reflecting their longer existence and probably more experience in technological progress. Companies founded between 1990 and 2004 have mixed maturity levels, with lower averages in the steel industry (1.98) and higher averages in the automotive industry (2.97), indicating variability in the adoption of Industry 4.0 technologies based on industry-specific factors. Recently established companies (2005–2020) generally show a higher level of maturity, especially in the automotive industry (3.06), suggesting a more modern approach and faster adoption of new technologies.

The analysis of the origin of capital shows that companies with foreign capital are characterized by a higher level of maturity, especially in the automotive industry (average value 3.72), which indicates significant investments and advanced technological capabilities. Mixed-capital companies show varying maturity levels, with lower maturity rates in the steel industry (1.89) and higher maturity rates in the automotive industry (3.30), reflecting a combination of opportunities and resources. Companies with purely Polish capital show a moderate level of maturity, ranging from 2.09 to 2.86, suggesting a steady but less pronounced implementation of Industry 4.0 technologies compared to their counterparts with foreign and mixed capital.

**Table 2**  Assessment of the maturity level of all pillars of Industry 4.0 according to the characteristics of the analyzed enterprises: Average values.

| *Elements* | | *Industry* | | |
|---|---|---|---|---|
| | | *Automotive* | *Steel* | *Food* |
| Headcount | 10–49 employees | 2.74 | 1.76 | 1.42 |
| | 50–249 employees | 3.12 | 2.86 | 2.72 |
| | Over 250 employees | 3.33 | 3.24 | 3.34 |
| Year of establishment | Until 1989 | 2.96 | 2.95 | 2.02 |
| | 1990–2004 | 2.97 | 2.32 | 2.25 |
| | 2005–2020 | 3.14 | 1.98 | 3.06 |
| Corporate capital | Polish | 2.86 | 2.09 | 2.18 |
| | Foreign | 3.72 | 2.06 | 1.89 |
| | Mixed | 3.27 | 3.48 | 3.30 |

*Source:* Own study.

One of the key conclusions is the significant variability in the adoption of Industry 4.0 technologies among Polish manufacturing companies. While some companies are at the forefront of digital transformation, using advanced technologies such as IoT, Big Data, and autonomous systems to optimize their operations, others are lagging behind due to various barriers.

This variability is attributed to factors such as organizational readiness, financial capabilities, and strategic vision. Companies that have successfully integrated these technologies are seeing improved efficiency, reduced costs, and enhanced product quality, demonstrating the tangible benefits of digital transformation.

The study identified several barriers to achieving a higher level of business maturity. The high upfront costs of implementing the technology are a significant challenge, especially for Small and Medium-sized Enterprises (SMEs). A lack of skilled labor proficient in new technologies also hampers progress, necessitating investment in training and development.

Resistance to change is another critical barrier. Organizational inertia and reluctance to rebuild traditional processes can slow down the adoption of new technologies. This resistance often stems from a lack of understanding of the benefits and fear of disrupting established workflows.

Successful digital transformation requires strategic planning and strong leadership. Companies that demonstrate a higher level of business maturity have leaders who are committed to digital innovation and who promote a culture of continuous improvement. These leaders play a critical role in formulating a clear vision, securing the necessary resources, and motivating employees to embrace change.

Strategic planning involves not only choosing the right technologies but also aligning them with the company's overall business goals. This ensures that technology investments contribute to strategic goals such as market expansion, customer satisfaction, and operational efficiency.

The results indicate a significant differentiation in the level of technological advancement of Polish manufacturing companies. While some companies are demonstrating advanced integration of Industry 4.0 technologies, others are still in the early stages of implementation. The key areas where technological advancements have been most evident are digital infrastructure, data management, and process automation.

Companies that have fully embraced Industry 4.0 technologies demonstrate a higher level of business maturity, characterized by optimized, standardized, and efficient processes. These companies are better equipped to achieve strategic and operational goals, stay competitive, and lead in the digital economy.

Despite the progress, the study also highlights several challenges businesses face in achieving full business maturity. Common obstacles include the high cost of implementing the technology, the lack of skilled labor, and resistance to change within the organization. In addition, the integration of advanced technologies such as IoT, Big Data, and autonomous systems remains a complex task that requires significant investment and strategic planning.

The study suggests that to meet these challenges, businesses need to focus on continuous improvement and innovation. This includes investing in employee training, fostering a culture of adaptability, and developing comprehensive digital transformation strategies. By addressing these areas, companies can enhance their technology capabilities and move to the next level of business maturity.

The findings of this study have important strategic implications for industrial companies, policymakers, and stakeholders in the digital economy. From a business perspective, the results highlight the importance of taking a systematic approach to digital transformation. This involves not only implementing advanced technologies, but also ensuring that processes and organizational structures are aligned with strategic goals.

The research presented in this book has several practical implications for the various stakeholders involved in the digital transformation of the manufacturing sector. These stakeholders include business leaders, policymakers, industry associations, technology providers, and educational institutions. The insights and recommendations in the study can help these stakeholders make informed decisions, develop effective strategies, and foster an environment conducive to the adoption of Industry 4.0.

Business leaders can use the results to prioritize strategic investments in Industry 4.0 technologies. By understanding the specific areas where technological advances can bring the highest returns, they can allocate resources more efficiently. This includes investing in IoT, Big Data Analytics, Automation, and Cybersecurity measures. The identified shortage of skilled labor highlights the need for continuous training and development of workers. Business leaders should invest in training programs to upskill their employees, ensuring that they are prepared to handle new technologies and processes. It can also be beneficial to work with educational institutions to develop specialized training modules.

Resistance to change is a common barrier to digital transformation. Leaders should focus on change management strategies to foster a culture of innovation and adaptability. This involves clearly communicating the benefits of new technologies, involving employees in the transformation process, and providing support during the transition. Research shows that a higher level of business maturity is associated with optimized and standardized processes. Business leaders can use this knowledge to evaluate and redesign existing processes, incorporating Industry 4.0 technologies to increase efficiency and reduce operating costs.

Policymakers can facilitate the adoption of digital technologies by developing supportive legislation and providing financial incentives. This includes grants, tax breaks, and subsidies for technology investments, especially for SMEs that may face high implementation costs. Investing in digital infrastructure, such as high-speed internet and secure data centers, is critical to enabling the widespread adoption of Industry 4.0 technologies. Policymakers should focus on building and upgrading the necessary infrastructure to support digital transformation across the manufacturing sector.

Governments can work with educational institutions and industry associations to develop comprehensive education and training programs. These programs should aim to equip the future workforce with digital competencies, bridging the current skill gap and ensuring a steady supply of skilled professionals. Encouraging public-private partnerships can stimulate innovation and accelerate the adoption of new technologies. Policymakers can create collaborative platforms between government agencies, businesses, and research institutions, fostering an ecosystem that supports technological advancement and knowledge sharing.

Industry associations can play a key role in promoting Industry 4.0 technologies and raising awareness of their benefits. They can host seminars, workshops, and conferences to educate their members on the latest technology trends and best practices for digital transformation. Associations can develop benchmarking tools and best practices to help their members evaluate progress and learn from successful implementations. By providing access to case studies and success stories, associations can inspire and guide companies on their digital transformation journey. Facilitating networking and cooperation between member companies can lead to the sharing of knowledge and resources. Industry associations can create forums and platforms where companies can collaborate on joint projects, share insights, and tackle common challenges together.

Technology providers can use the insights from the study to develop solutions tailored to the specific needs of Polish manufacturing companies. Understanding the unique challenges and requirements of this market enables suppliers to offer more relevant and effective products and services. Suppliers should offer their customers comprehensive support and training, ensuring that they can fully leverage the capabilities of new technologies. This includes not only technical support but also training programs that help customers integrate and optimize technologies in their business. The research indicates areas where technological progress is most needed. Technology providers can focus their innovation efforts on these areas, developing new products and solutions that fill the identified gaps and increase the overall digital maturity of manufacturing companies.

Educational institutions can revise their curricula to include courses on Industry 4.0 technologies and digital transformation. By equipping students with the necessary skills and knowledge, these institutions can provide a steady supply of qualified professionals ready to enter the workforce. Universities and research institutions can collaborate with industry on research projects that explore new applications of Industry 4.0 technologies. This collaboration can lead to innovative solutions and provide valuable insights that drive further advances in the field. Offering lifelong learning opportunities and continuing education programs can help current professionals stay up-to-date on the latest technological developments. Educational institutions can provide flexible learning options, such as online courses and workshops, to accommodate the needs of working professionals.

Based on the findings, the study provides several recommendations for enterprises looking to increase their business maturity:

1. **Invest in technology and training:** Companies should devote resources to acquiring advanced technologies and training their employees. This dual investment ensures that technology tools are used effectively and that employees are able to manage and optimize these tools.

2. **Foster a culture of innovation:** Creating an organizational culture that values innovation and is open to change is key. Encouraging experimentation, rewarding innovative ideas, and creating cross-functional teams can drive continuous improvement and adaptation.

3. **Develop strategic partnerships:** Collaboration with technology providers, research institutions, and other industry players can provide access to new technologies, insights, and best practices. These partnerships can also facilitate the sharing of knowledge and resources, making the digital transformation process more efficient.

4. **Leverage data analytics:** Using data analytics to gain insights into operations, customer behavior, and market trends can help businesses make informed decisions. Data-driven decision-making enables companies to identify opportunities for optimization and innovation.

5. **Strengthen cybersecurity measures:** As businesses adopt more digital technologies, ensuring robust cybersecurity becomes a necessity. Protecting digital assets and data from cyber threats is essential to maintaining operational integrity and customer trust.

The research presented in this book allows for a nuanced understanding of the business maturity of Polish manufacturing companies in the era of Industry 4.0. The results highlight the critical role of technology adoption in driving business growth and competitiveness. By addressing the identified barriers and implementing recommended strategies, enterprises can navigate the challenges of digital transformation and achieve sustainable success.

This comprehensive analysis provides a valuable resource for business leaders, policymakers, and researchers interested in the dynamics of technological progress and its impact on industrial companies. It highlights the importance of a strategic, well-coordinated approach to digital transformation, emphasizing the potential for significant gains in efficiency, productivity, and market competitiveness.

# Index